How F.U.N.K.Y. Is Your Phone?

Over 300 Practical Ways to Use Your Cell Phone

Functional Useful Nuggets of Knowledge Tailored Specifically For You

written by denise barnes
designed by rob ripley
foreword by brian underdahl

I0462933

iKeep It Funky™ "How F.U.N.K.Y. is Your Phone?"

This edition was printed by Create Space.

For information, you may contact iKeep It Funky at:
iKeep It Funky
Tel: 206-659-8239
E-Mail: Info@iKeepItFunky.com
Website: http://www.iKeepItFunky.com

ISBN 1449586759
EAN-13: 9781449586751

10 9 8 7 6 5 4 3 2

Layout And Design: Rob Ripley

Technical Editor: StormKatt Productions

Poem Contributions: Mélanie Hope And Marrico Gordon

Manufactured in the United States Of America

DISCLAIMER:

The information contained in the printed, digital and/or audio versions of this book is strictly for entertainment purposes. The author/publisher used best efforts in preparing these materials with the understanding that technology is volatile and some resources may change after publishing. "iKeep It Funky" makes no guarantee of the accuracy, completeness, usefulness or adequacy of any resources, information, apparatus, product, or process available from external sites listed and proposes no advice as such. As always, seek the advice of competent professionals before making any financial, legal or medical choices.

Please Note: The publisher and associated artists state that they used the names of trademarked entities throughout this book only for editorial purposes and to the benefit of the trademark owner with no intention of infringing upon that trademark. All information provided is of a general nature and is not intended to address the circumstances of any particular individual or entity.

ACKNOWLEDGMENTS

I am truly blessed to have an awesome circle of support and mentorship. I will not attempt to list every person that has in some manner or fashion contributed in the preparation of this book. It would take pages to list every individual. Here is a short list of individuals that have been instrumental in this project.

Contributions include (listed alphabetically):

Allison Ruggles

Barbara Honey

Brianna Baptiste

Carolina Cruz

Cruz Rodriguez

David Grigsby

Debrena Jackson Gandy

DJ Topspin AKA Barry Gayle

Don R. & Janet Crawley

Donna Warren

Geoff Santos

Janice & Harry Barnes

Margue Hunt-Familton

Marrico Gordon

Natasha Smith

Rebecca Little

Shannon Evans

Tami Smith

My graphic designer Rob Ripley and editor Mélanie Hope have been a God send!! Thank you for your dedication, effort and creativity!

Dedicated to my Creator without whom not even my existence would be possible.

FOREWORD

Some years ago no one gave any thought to owning their own computer. After all, who would have room for one of those giant, room filling monsters—much less any use for one? My, how things have changed! Today almost everyone has a powerful computer in their pocket, purse, or on their belt. The only thing is, now we call them Smartphones.

True, we mostly don't think of our Smartphones as computers, but that's really what they are. With a modern Smartphone you can surf the Web, exchange e-mails, swap text messages, play games, work on a spreadsheet, navigate to new destinations, and, oh yes, make phone calls.

Today's Smartphones represent a blending of devices. When I wrote Pocket PCs For Dummies a bit over ten years ago, there wasn't any crossover between mobile phones and PDAs (personal digital assistants). You used your mobile phone to make phone calls and that's all. If you wanted something to help organize your life you got a Palmpilot or a Pocket PC. Mobile Web surfing, on-the-go e-mail, and such were pretty much out of the question. Eventually, though, people started to realize that you could combine the mobile phone and the PDA, and the result was today's ubiquitous Smartphone—the go anywhere, do almost anything device.

While the Smartphone may be more powerful than most personal computers of just a few years ago, it is not simply a replacement for a PC. Rather, the Smartphone is a new type of computing device with its own unique qualities and capabilities. That's why a book like this one from Denise Barnes is so useful. Denise has compiled a set of very useful information which will help you make the most of your Smartphone. Here you'll find things you didn't know, things that will make you laugh, and things that will make you think. Who could ask for more?

Brian Underdahl, Author

www.underdahl.net

"The Smartphone is a new type of computing device with its own unique qualities and capabilities. That's why a book like this one from Denise Barnes is so useful."

~Brian Underdahl -Author

Denise Barnes, Founder of iKeep It Funky, is known as 'fashionably geeky' as she unapologetically and enthusiastically loves anything related to gadgetry and technology. Her career of developing devices, embedded and mobile platforms and mobile applications for the world's largest software manufacturer surged her geeky passion and made her a trusted advisor for mobile technology. Besides being a techno-fashionista, Denise is a globe-trotter who enjoys reading, music, walking and fast cars.

In a recent interview, Denise shared how her passion for mobile technology began:
"My love for mobile technology resonated when I was preparing for a business trip to China. I had no time to learn the native language - Mandarin. I needed to get around the city in a timely manner for my customer meetings. Of course, I also wanted to do some site seeing. At this time I was working with quite a few independent software vendors in bringing to market some mobile apps. I purchased and installed a mobile translation app. This software translated English to Mandarin in text. It also provided the translation in the native tongue and an associated sound using the local accent. It even showed pictures.⊠ I was able to get around to and from the hotel, the convention center where I was attending a mobile conference, and even communicate with a taxi driver.

I have always been one that was very technically inclined and loved technology, but I absolutely fell in love with mobile technology during that trip. I then led numerous mobile technology communities and assisted in the development of several mobile platforms and mobile devices. There is not any task I can think of where mobile technology cannot assist."

Denise's experience led her to create iKeep It Funky with the following mission statement:
It is the mission of iKeep It Funky to improve the lives of others through the advanced use of mobile technology for greater efficiency and productivity.

Rob Ripley is a graphic designer and photographer that lives in Seattle, Washington. Rob graduated from Pacific Lutheran University with a Bachelor of Fine art degree focusing on visual communication and pho-tography. Rob spends his free time taking photos, creating art and doing freelance graphic design for print and the web; also, he runs his own art business known as Artifakt. Artifakt is a multi-genre art show fusing music and art together creating a unique art going experience.

"A world of mobile technological growth alone is not beneficial without each and every individual's personal growth!"
~*Denise Barnes -Author*

TABLE OF CONTENTS

PART 3 ~   **ON THE GO FREEDOM: MOBILE PRODUCTIVITY**

ADAM'S PROGRESS

by Marrico Gordon

OMG

LOL

TEXT MESSAGING ON THE BUS.

BLONDE WITH PINK GUCCI FRAME PRESCRIPTION GLASSES

SITS AT A GREEN LIGHT

TEXT MESSAGING AS THE DRIVER BEHIND HER

HONKS TWICE.

INTERNET SURFING KIDS AT THE PUBLIC LIBRARY

WILL KEEP YOU WAITING.

PLAYING GAMES, GOSSIPING ON MYSPACE,

DOING EVERYTHING BUT LEARNING.

RESUMES SENT VIA EMAIL

JOB APPLICATIONS DONE ONLINE.

ONLINE COURSES,

ONLINE DATING,

ONLINE SHOPPING.

ILLUMINATING SCREENS

HYPNOTIZING BEINGS.

TTYL,

SHE HAS A BUSINESS MEETING AT 9,

VIA SATELLITE,

ON HOW TO CREATE A WEBSITE.

STUDENTS RESEARCHING ONLINE,

ARE DISTRACTED BY POP-UP INSTANT MESSAGES ON FACEBOOK.

"NEW WINDOWS 7 SPLIT SCREEN FEATURE, SO COOL."

"DOWNLOADING LIL' WAYNE'S LATEST VIDEO, SO COOL."

"THE IPHONE, SO COOL."

"IPOD NANO, SO COOL."

"DID YOU SEE THE PANTS ON THE GROUND VIDEO ON YOUTUBE?"

"THE WAY THAT OLD MAN SANG AND DANCED, LOOKIN' LIKE A FOOL."

FROM RUBBING STICKS TO KEROSENE LAMPS TO

WIRELESS FLAT SCREEN TELEVISIONS.

Fire, fire everywhere.

Solar powered cars!

Zoom zoom.

The speed of livin', zoom zoom.

A teacher can teach without being in the classroom.

LMAO on my bluetooth.

Parents asking their children

How to save numbers in their new phone.

"I don't know. Ask your granddaughter,"

Who is only 9 years old.

Enhanced features, lost weight, new nose.

Procedures that cost millions,

Foods cloned.

Stem cell research and genetically altered babies!

It's a brave new world.

Obama kept his Blackberry?

What a brave new world.

You can be playing Madden10 against some kid

Clear across the coast.

Social stability, coordinated activities.

The internet's a vehicle.

Technological modifications,

Applications and infatuations with devices

That make things a little easier.

Like the remote.

Thank man for DVR,

Recording every episode of Heroes.

Less cumbersome activities,

Mapquesting our way through every road.

Adam's progress from microorganism to microchip.

PART 1

WHAT MOST PHONES DO AND WAYS TO DO IT BETTER...

- ASSISTIVE TECHNOLOGY
- DATA RECOVERY
- SOCIAL MEDIA

Social Networking

Current Events

A blind man once asked a wise man:
"Can there be anything worse than losing your eye sight?"
The wise man replied:
"Yes..losing your 'vision'.."
~A Blind Man

ASSISTIVE TECHNOLOGY

ASSISTIVE TECHNOLOGY
AUGMENTED PHONES AND APPS FOR THE DISABLED

THE CLARITYLIFE AMPLIFIED GSM MOBILE PHONES
This cool cell phone is twice as loud as ordinary cell phones and has larger buttons. There is a flashing visual ringer to signal a call, built in flashlight and other cool features. You can get a SIM Card from AT&T or T-Mobile.

VLINGO
Voice recognition to send texts, emails, phone calls, browse the web, update social networking sites, etc.
Applicable Devices: iPhone, Windows Mobile, Nokia, Blackberry
http://www.vlingo.com

NUANCE TALKS
Converts text on a cell phone into audio including Caller ID.
Applicable Devices: Nokia And Samsung
http://www.nuance.com/talks/downloads.asp

K–NFB READING TECHNOLOGY
Solutions for the blind such as text-to-speech by a single activation button and many more solutions.
http://www.knfbreader.com/index.php
TELEPHONE: 1-877-547-1500

F.U.N.K.Y. TIP:
Cool Senior resources can be found at http://www.savvysenior.org seniorresources.htm#ASD.

DRAGON

Use your voice to search for a contact, compose an email and much more!

http://www.dragonmobileapps.com
TWITTER - http://twitter.com/dragontweets
FACEBOOK - http://www.facebook.com/dragonmobileapps

JITTERBUG

Cheap cell phone plans as low as $14.99 a month. No contracts! You can also add helpful services to your gift such as roadside assistance and more. The cool feature of these cell phones is that you can select YES or NO for options instead of navigating through a bunch of menus.
Jitterbug is a no-frills (no texting, low minutes) phone that is great for seniors.
http://www.jitterbug.com
TELEPHONE 1-800-733-6632

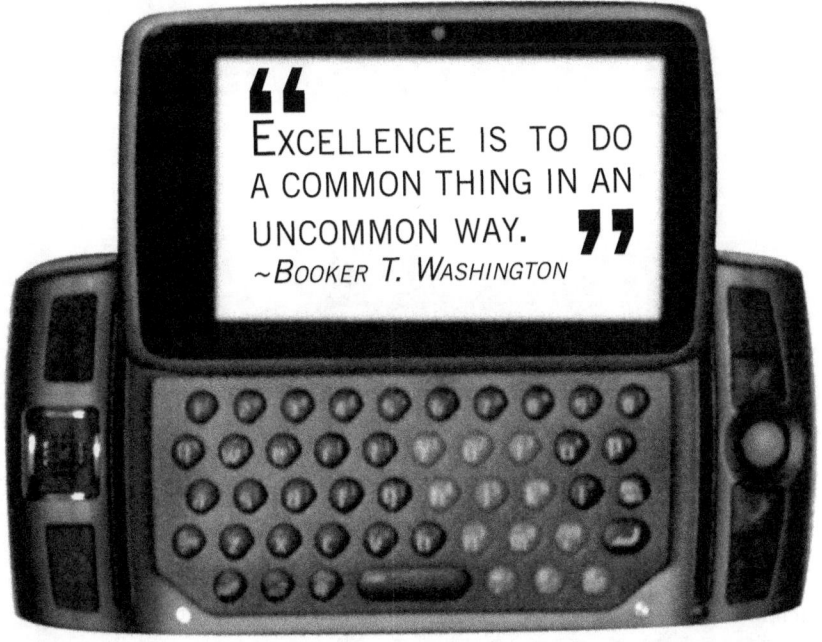

EXCELLENCE IS TO DO A COMMON THING IN AN UNCOMMON WAY.
~BOOKER T. WASHINGTON

"Lack of planning on your part, does not justify an emergency on our part."
~ *Unknown*

DATA RECOVERY

DATA RECOVERY
Sync backup and recover your phone's data

CELLEBRITE
Transfers your cell phone content from one cell phone to another.
http://www.cellebrite.com

GOOSYNC
Easy way to sync Gmail, Google Calendar, Contacts and tasks on your mobile device.
Applicable Devices: iPhone, Windows Mobile, Blackberry, Nokia, Palm
http://www.goosync.com

SAVECELL
Never lose your cell phone contacts.
Applicable Devices: Nokia, more
http://savecell.co.uk

SYNCMATE
Sync your Windows Mobile, USB flash drives, PlayStation Portable (PSP), and Google accounts and back up your data all online.
This app allows you to sync multiple devices.
http://www.sync-mac.com

SYPHONE
Back up your text messages.
Applicable Device: iPhone
http://syphone.selcukyilmaz.com

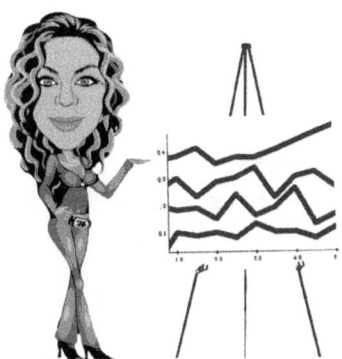

F.U.N.K.Y. STATISTIC:

Mobile technology allows users to post instant and frequent updates via their mobile phones, and is growing exponentially this year to an estimated 18 million users a month - a 200% increase over 2008.

"Always remember you're unique, just like everyone else."
~Unknown

SOCIAL MEDIA

SOCIAL NETWORKING

DON'T BE A TWIT, STAY CONNECTED TO YOUR FRIENDS NO MATTER WHERE YOU OR THEY MAY BE

DAILYMOTION
Capture video and share with your family and friends, search millions of other videos and more.
http://www.dailymotion.com/sas/iphone
WAP (Mobile WEBSITE) http://m.dailymotion.com

FLICKR MOBILE
Share your photos with anyone.
WAP (Mobile WEBSITE) http://m.flickr.com

MOBILICIO.US
Access Delicious bookmarks from your mobile device. (formerly called del.icio.us).
http://mobilicio.us

NETLINGO
List of text & chat acronyms for those of you wanting to learn to be more tech and net savvy.
Applicable Device: iPhone
http://www.netlingo.com/iphone
http://netlingo.com/acronyms.php

RAINEDOUT
For you coaches, soccer moms, or any organization that you may manage - use this service for communications of cancellations, scores, or changes to any announcements. You may wish to deliver information to your teams or members via SMS.
http://www.rainedout.com

SMARTSYNC
Keep your iPhone contacts up-to-date with Facebook contact profiles, etc.
Applicable Device: iPhone
http://www.ultimake.com/smartsync

TIPS TO TEXTING
Dating advice.
http://www.tipstotexting.com

VOTE ON A BILLBOARD
Cool billboard voting in New York via SMS.
http://www.funsms.net
http://www.funsms.net/sms_in_marketing.htm

EPLIXO
Create a Video Chat Party.
http://www.eplixo.com/blog
http://eplixo.com/m

YOU IN?
Use this app to communicate your good deeds for others to join you, inspire and motive random acts of kindness.
WAP (Mobile WEBSITE) http://m.yahoo.com

EASEMS
Send SMS messages from your PC using your MobileMeWorld account.
http://www.mobilemeworld.com/pages/products/easemsconsole.php

FISHTEXT
Send cheap SMS text to over 200 countries.
http://www.fishtext.com

FRING
Live talk, chat, and instant message.
Applicable Devices: iPhone, Nokia, Android
http://www.fring.com

GIZMO SMS
Send SMS messages via this web site.
http://www.gizmosms.com

F.U.N.K.Y. TIP:
Sitting to close to a television will fatigue a person's eyes, yet will not permanently damage them. The same is true if reading in low light.

SOCIALTEXT MOBILE

Share activity, stay abreast of conversations in microblogging sites, and collaborate with customers or coworkers on the go.
Applicable Devices: iPhone, Blackberry
http://www.socialtext.com/products/mobile.php

TXTDROP

Send text messages from your computer with cool widgets.
http://www.txtdrop.com

WAY2SMS

Send SMS to a cell phone, receive email alerts for Yahoo or Gmail, and chat with Gtalk and Yahoo contacts.
http://wwwf.way2sms.com/content/index.html

F.U.N.K.Y. LAW:
Currently, you can use your cell phone in traditional methods (wired) for an emergency situations only. This is soon to change.

LOOPT

Be 'loopt' into events and happenings around you.
Applicable Devices: Blackberry, iPhone, Android
http://www.loopt.com

POPURLS

An aggregate of all the up-to-the-minute headlines from the most popular news, blogs, vlogs, etc.
Applicable Devices: iPhone, Android
http://popurls.mobi
http://i.popurls.mobi (for iPhone And Android)
http://b.popurls.mobi (for Blackberry)

TWITTER

Get top stories from an aggregate of popurls.
TWITTER http://twitter.com/popurls

QIK

Live real-time media video streaming to the internet.
Applicable Devices: iPhone, Blackberry, Windows Mobile, Nokia, Palm, Android
http://qik.com

WARNING: DATES IN CALENDAR ARE CLOSER THAN THEY APPEAR.
~UNKNOWN

WCCO MOBILE LOCAL NEWS

Obtain updated local news.
Applicable Devices: iPhone, Android
http://wcco.com/wireless

Sign Up For SMS Alerts: Weather (every day Mon - Fri) TEXT the word "WCCO" to "66247"

Flu Alerts: TEXT the word "flu" to "66247" for seasonal flu information.

Minnesota Vikings Alerts: TEXT the word "VIKINGS" to number "66247"

Receive deals: TEXT the word "DEALS" to number "66247"
WAP (Mobile WEBSITE) http://m.wcco.com

CHACHA

Search a question, get an answer.
http://www.chacha.com
Text "DOH" to "242242"

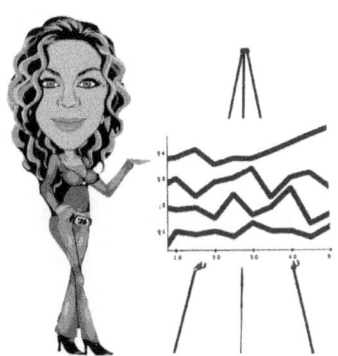

F.U.N.K.Y. STATISTIC:

In 2009, Twitter became the mode of communication for any artist, celebrity, politician, grandma/grandpa, to schoolgirls who wished to display cutting edge information.

COOL QUOTES
Obtain cool quotes.
Applicable Device: iPhone
http://itunes.apple.com/in/app/cool-quotes/id343051586?mt=8

WINK
Create photostrips to publish on Facebook, Flickr, or Twitter.
Includes tools to resize or add effects to your photos.
Applicable Device: iPhone
http://wink.shutterfly.com

F.U.N.K.Y. TIP:
You can save money by dialing (800) FREE 411, or (800) 373-3411, instead of dialing 411.

CURRENT EVENTS
STAY UP TO DATE

MOBILE-FINANCIAL.COM
Mobile financial news.
http://mobile-financial.com

THE DAILY JOURNAL TEXT ALERTS
Receive customized SMS alerts for news, sports stats, stocks, etc.
http://www.thedailyjournal.com/sms

WALL STREET NEWS
Wall Street alerts.
This free service can be discontinued at any time by replying to any one of the alerts with the word "stop".
http://www.WallStreetNewsAlert.com
Sign up for Wall Street Alerts via SMS for news or stocks.
TEXT the word "press" to "68494"
TWITTER http://twitter.com/wsna

ELECTION UPDATES
Receive SMS news about elections.
http://www.rnc.org

KYODO NEWS
Japanese news.
Applicable Devices: iPhone, iPOD Touch
http://kijizo.jp/iphone/index_e.html

PART 2

AMAZING RESOURCES YOU CAN ACCESS MOBILY...

- **ENTERTAINMENT**
  - *Literature*
  - *Music*
  - *Video*
  - *Sports*
  - *Games*

- **SHOPPING**
  - *Barcode*
  - *Phone Service*
  - *Catalogs*
  - *Coupons & Sales*
  - *Auctions*
  - *Food*
  - *Real Estate*
  - *Automobiles*
  - *Specialty*
  - *Resources*

- **BETTER WORLD**
  - *Benevolence*
  - *Going Green*

- **WELLNESS**
  - *Crisis*
  - *Medical*
  - *Nutrition*
  - *Spiritual*
  - *Self Help*

- **PARENTING**
  - *Resources*
  - *Monitoring*
  - *Alerts*
  - *Fun*

- **BONUS**
  - *Fun*

"Music exalts each joy, allays each grief, expels diseases, softens every pain, subdues the rage of poison, and the plague."
~John Armstrong

ENTERTAINMENT

LITERATURE
DOWNLOAD BOOKS AND MAGAZINES RIGHT TO YOUR PHONE

NOOK
Read books from Barnes & Noble electronically.
Yes, you can read your eBooks across a wide range of devices in addition to your Nook. To do this, simply download the free Barnes & Noble eReader app on your iPhone, iPod Touch, BlackBerry smartphone, PC or Mac computer. All of your Barnes & Noble content will be available on those devices too.
Applicable Devices: Blackberry, iPhone
http://www.barnesandnoble.com/ebooks/download-reader.asp

RANDOM HOUSE
A collection of books and audio CDs Sent via SMS.
http://www.rbooks.co.uk
TWITTERhttp://twitter.com/randomdigital

READ IT LATER
Read websites offline.
Applicable Devices: iPhone, Android
http://readitlaterlist.com
WAP (Mobile WEBSITE) http://readitlaterlist.com/l

TIME MOBILE
Time Magazine.
Applicable Devices: iPhone, Blackberry
http://app.time.com
WAP (Mobile WEBSITE) http://mobile.time.com

MUSIC

Music lovers can enjoy tunes in a greater way

BRITNEY.COM
Britney Spears ringtones, news alerts, etc.
http://www.britney.com/us/mobile

DJ APP
DJ mixer pro app.
Applicable Devices: iPhone, iPOD Touch
http://djmixer.fm

KELLYOFFICIAL.COM
Kelly Clarkson ringtones.
http://www.kellyofficial.com/us/mobile

MEDIA MONKEY
Alternative to syncing music on your iPhone or iPod without using iTunes.
http://www.mediamonkey.com/download

METROLYRICS
Find lyrics to a song from your cell phone.
http://www.metrolyrics.com
WAP (Mobile WEBSITE) http://m.metrolyrics.com

F.U.N.K.Y. LAW:
When a music CD is purchased, the purchaser may have the technological ability to rip a CD into another digital format such as an mp3 or other digital file, often for use on portable music players.

MICHAEL JACKSON MOBILE APPLICATIONS
All things Michael Jackson.
http://www.michaeljackson.com/us/mobile

MOBILETUNES
Awesome tool for music artists, bands and record labels to bring promotions and advertising to their fans with mobile music channel that provides the latest information about trends and fan page information.
Can be obtained via SMS.
http://www.mobiletunesdj.com

MYPLAY
Music downloads and ringtones.
http://mobile.myplay.com

OFFICIAL CIARA MOBILE SITE
Music, ringtones and apps - all Ciara.
http://www.ciaraworld.com/us/mobile

PANDORA
Streaming internet radio based on personal preferences you define. Gracenotes lyrics service also now provided.
Applicable Devices: Android, Blackberry, iPhone, Palm, Windows Mobile
http://www.pandora.com/on-the-go

V CAST MUSIC WITH RHAPSODY
Subscribers can stream songs from their mobile devices.
Applicable Devices: iPhone, Android, Blackberry
http://www.rhapsody.com/electronics/phoneapps

SEND FREE RINGTONES

Upload ringtones for free, sent to your phone instantly.
http://www.ringerdrop.com/?ref=txtdrop

SUPER RADIO

Internet streaming player.
Create preset channels etc.
http://wareous.com/superradio

TUNEROOM

Music search engine.
Search and download as much music as you like.
WAP (Mobile WEBSITE) http://www.sideload.com/m/index

WANNA PLAY

Learn to be a DJ or to play an instrument.
Applicable Device: iPhone
*http://itunes.apple.com/WebObjects/MZStore.woa/wa/
viewSoftware?id=314865990&mt=8*

VIDEO

MAKE, SHARE AND SEARCH VIDEOS USING THESE APPS

DIVX MOBILE PLAYER
Video viewing application.
Applicable Device: Windows Mobile Phone
http://labs.divx.com/MobileDownload

EYETV 3.3
iPhone app for video streaming.
Applicable Device: iPhone
https://live3g.eyetv.com

EYETV LIVE3G
Stream live television and EyeTV recordings anywhere.
Applicable Device: iPhone
https://live3g.eyetv.com

F.U.N.K.Y. LAW:
Anyone in a public place can take pictures of anything. Public places include parks, sidewalks & malls.

SPORTS

WATCH YOUR FAVORITE TEAMS AND GET
UP-TO-THE-MINUTE SCORES OR IMPROVE YOUR GOLF SWING

BETTINGEXPERT
Sign up to receive SMS tips for betting odds and expert tips for betting Soccer, Football, other sports including Poker.
http://www.bettingexpert.com
TWITTER @bettingexpert - http://twitter.com/bettingexpert
FACEBOOK http://www.facebook.com/bettingexpert

MOBILE SPORTS SCORES
WAP sports scores.
WAP (Mobile WEBSITE) http://m.yahoo.com/sports

GOLF
Find a golf course right from your mobile device (US Only).
Applicable Devices: iPhone, iPod Touch, Blackberry, Windows Mobile, Palm
http://mobile.zagat.com/iphone_golf.htm

INSTA PRO
Golf Analysis Tool. Compare your swings.
Applicable Device: iPhone
http://theinstapro.com/home

NFL ALERTS
Receive NFL Playoff alerts and more.
http://www.nfl.com/mobile

GAMES
BONUS GAMES RIGHT IN YOUR HAND
(SEE VERSION TWO FOR MUCH MORE)

BULLY PIX GAMES
Download a variety of games.
Applicable Device: iPhone
http://www.bulkypix.com
http://www.bulkypix.com/jeux/presentation/magnetic-sports-soccer

GAMEHOUSE
Download games for your iPhone or iPod Touch.
Applicable Devices: iPhone, iPod Touch
http://www.gamehouse.com/iphone-games

GAMEHOUSE'S COLLAPSE
Download the Collapse game.
Applicable Devices: Android, BlackBerry, iPhone
http://www.collapse-games.com/default.htm

HOTEL MOGUL
Game to help Lynette find her cheating husband.
Applicable Device: iPhone
http://iphone.alawar.com/?p=231

TAP TAP REVENGE
Music game of top ten artists with lighting and animation effects.
Applicable Device: iPhone
http://tapulous.com/dance

MOBILE GAMERTAG V2.3

AMNYA from x360me.com has released version 1.1 of the world's first free Xbox 360 Gamercard application for Windows Mobile. Now, users of a PDA with a VGA screen can get their gamercard directly on their windows mobile phone.
http://x360me.com

SHERLOCK HOLMES: THE GAMES IS AFOOT

Applicable Device: iPhone
http://www.mobiledeluxe.com

SMS GAMERS TIPS

Get strategies, tips, hints and cheats via SMS for XBox, Playstation, And WII.
http://sms4gamecheats.net/rts.html

SPEED TEXTING GAME

See where you rank.
http://mobiledeluxe.com/gs-gaming/games/speed-texting.html

"The safest way to double your money is to fold it over once and put it in your pocket."
~Kin Hubbard

SHOPPING

BARCODE

SCAN, SEARCH AND COMPARE BARCODES DIRECTLY FROM YOUR PALM

ANDROID BARCODE SCANNER
Barcode reader.
Applicable Device: Android
http://google.com/m/products
http://www.google.com/support/mobile/bin/answer
py?hl=en&answer=167247

BEST BUY ON DEMAND PRODUCT INFORMATION
TEXT the product keyword "SKU" to "332211" (obtain product keywords SKUS in Best Buy weekly inserts or in Best Buy stores). You will be sent an SMS with product details: name, make, model, color, rating and more. You will also receive the link to the product page on Best Buy's mobile site.
http://m.bestbuy.com

I-NIGMA BARCODE READER
Barcode reader.
Applicable Devices: iPhone, Blackberry, Windows Mobile, Nokia, Palm, Android
http://www.i-nigma.mobi

QR CODE / 2D BARCODE READER APP

Finger-friendly personal finance app for Windows Mobile.
Applicable Device: Windows Mobile Phone
http://winmoappstore.com

QUICKMARK

Mobile barcode reader.
Applicable Devices: iPhone, Nokia, Windows Mobile
http://quickmark.cn/En

BIG BAR, LITTLE BAR, BIG BAR, BIG BAR...
~UNKNOWN

PHONE SERVICE

CHEAPER RATES FOR LONG DISTANCE, CELL PHONE
PLANS FOR LESS...AND MORE

PENNYTALK
International calling service at $.02 a minute.
Applicable Devices: iPhone (Blackberry coming soon)
http://www.pennytalk.com

SKYPE
Make those long distance calls at a much cheaper rate by using.
VoIP (Voice Over IP).
Applicable Devices: iPhone, Windows Mobile
http://www.skype.com/mobile

PHONES4LIFE.COM
Provides free emergency services for seniors over 60 years old.
http://www.aboutus.org/Phones4Life.com

F.U.N.K.Y. TIP:

When shopping for a new cell phone, find a cell phone with a low SAR. To view a list of cell phones with low SAR levels go to http://budurl.com/lowsarcellphones.

CATALOGS

ONLINE CATALOGS RIGHT ON YOUR PHONE

IKEA

IKEA's 2010 Catalogue.
Applicable Devices: iPhone
iTunes Store Download; http://onlinecatalogue.ikea.com/2010/ikea_catalogue/gb/iphone

AMAZON

Search Amazon items in your palm.
Applicable Devices: iPhone and Blackberry
Blackberry - http://www.amazon.com/gp/anywhere/sms/bbapp
iPhone - http://www.amazon.com/gp/feature.html?docId=1000291661
SMS - http://www.amazon.com/gp/anywhere/sms

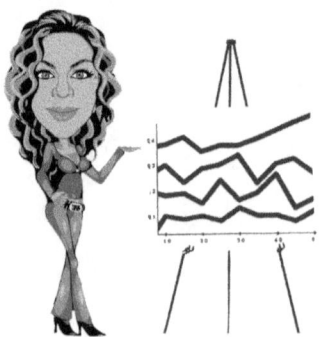

F.U.N.K.Y. STATISTIC:
According to a recent ACI Study(2),
18% of consumers have been victims
of credit or debit card fraud in the past
five years.

COUPONS & SALES

FIND ONLINE AND LIVE SALES DISCOUNTS AND COUPONS FOR YOUR FAVORITE STORES

BEST BUY REWARD ZONE GAMERS CLUB MOBILE ALERTS
Be the first to know about new releases and offers for video games.
http://m.bestbuy.com

CELLFIRE
Coupons via SMS.
http://www.cellfire.com

GASBUDDY
Cheapest gas alerts.
http://GasBuddyToGo.com

MIAPP
Drag and drop interactive ads to multiple devices.
Applicable Devices: iPhone, Blackberry, Android, Nokia
http://www.goldspotmedia.com

QPINS.COM
SMS alerts for deals in the Northwest for recreation, health & beauty, services, community.
http://www.qpins.com

SHOPTEXT
Get electronic coupons from your favorite brands.
http://www.shoptext.com
For Demo: TEXT "ShopText" to "467467"

SHORTCUTS
Get coupons from popular grocery stores such as Safeway and QFC.
http://www.shortcuts.com
TWITTER http://twitter.com/Shortcuts

SLICKDEALS.NET
Save money and find the lowest cheapest price, best deals and bargains, along with coupons. Also, find discount codes, promo codes, reviews and price comparisons.
WAP (Mobile WEBSITE) http://m.slickdeals.net
TWITTER http://twitter.com/slickdeals

VALPAK
Get coupons in your area listed by zip code.
(Soon Google's Android and Palm Pre apps will be added.)
Applicable Devices: iPhone
http://www.valpak.com/coupons

F.U.N.K.Y. TIP:
Did you know that you can get stylish bowling shoes for .85 cents at the bowling alley.

AUCTIONS

FROM EBAY TO AMAZON: BUY AND SELL GOODS ON THE GO

CEX
Buy or sell a used cell phone.
WAP (Mobile WEBSITE) http://m.webuy.com

CHRISTIE'S FINE ART, CONTEMPORARY PAINTINGS AND MORE
Auction events are sent via SMS. You can obtain art by Picasso,
Van Gogh and more. This is a place for both novices and connoisseurs.
http://www.christies.com

EBAY
Bid on an item from your mobile device.
http://pages.ebay.co.uk/mobile

MOBSAVER
Shopping price comparison for Amazon and Ebay.
http://www.mobsaver.com

SAW IT...WANTED IT...HAD A FIT...
GOT IT!
~UNKNOWN

FOOD

FIND THE BEST RESTAURANTS, GET REVIEWS, EVEN ORDER TAKEOUT VIA SMS

CAMPUSFOOD
Place orders via SMS.
http://www.campusfood.com

GROCERY IQ
Stay in touch with your favorite retailers, including Kroger, Safeway, CVS, Walgreens, Kmart and hundreds more. Features include barcode scanning, coupon integration and list sharing with multiple users.
Look for Coupons.com in the iTunes Store
Applicable Devices: iPhone, iPOD

ALLERGY ALERTS
FSA will send alerts regarding missing information from food labels that are missing or incorrect.
http://www.food.gov.uk/safereating/allergyintol/alert
TEXT "message "START ALLERGY" to the number "62372"

PIZZA HUT
Order Pizza from your cell phone.
WAP (Mobile WEBSITE) www.pizzahut.com/mobileordering

SUBWAY NOW
Order a Subway sandwich directly via text message.
Send a TEXT to "466626" for a list of items.
Reply back with the item you wish to order and receive a confirmation directly.
http://www.subwaynow.com

ZAGAT
Get ratings for restaurants, nightspots, hotels & shops.
You can even make reservations.
Applicable Devices: Android, Blackberry, Windows Mobile, iPhone, Palm
WAP (Mobile WEBSITE) *ZAGAT.mobi*
http://www.zagat.com/mobile

REAL ESTATE

RESEARCH LOCAL MARKETS OR FIND YOUR NEXT HOME
USING YOUR MOBILE PHONE

HOMES ON MOBILE PHONES
Home listing SMS alerts.
http://www.homesonmobilephones.com/web/en/index.cfm
TWITTER http://twitter.com/Homes_on_Phones

ICODE
Home buyer's tool to enter code from sale sign into mobile phone.
http://www.clearskymobilemedia.com

NEW YORK TIMES REALESTATE
Search properties in New York using listing ID.
http://www.nytimes.com/realestate
WAP (Mobile WEBSITE) m.nytimes.com/re

TXT2LOOK
Home buyers can text a code on particular signs to receive
property listing details.
http://www.txt2look.com

AUTOMOBILES

RESEARCH YOUR NEXT RIDE OR GET TIPS ON MAINTENANCE AND REPAIR

CARTOPIA
Be informed when shopping for a new car.
http://www.nationwide.com/mobile/cartopia.jsp
TWITTER http://twitter.com/nationwide
FACEBOOK http://www.facebook.com/nationwide

KELLY BLUE BOOK
Get new or used car details right from your iPhone.
http://www.kbb.com/kbb/CompanyInfo/Mobile.aspx

MINI UK
Receive an SMS OR MMS alert to find your perfect mini van.
http://www.minicherished.co.uk

NEW CAR GUIDE
Guide for new cars that are or will be sold in the UK in 2010.
Download via iTunes:
http://itunes.apple.com/gb/app/new-car-guide-uk-2010/id344255527?mt=8&uo=6

F.U.N.K.Y. LAW:
A jurisdiction-wide ban on driving while talking on a hand-held cellphone is in place in 7 states (California, Connecticut, New Jersey, New York, Oregon, Utah, and Washington) and the District of Columbia.

Text messaging is banned for all drivers in 20 states and the District of Columbia. In addition, novice drivers are banned from texting in 9 states (Delaware, Indiana, Kansas, Maine, Mississippi, Missouri, Nebraska, Texas, and West Virginia) and school bus drivers are banned from text messaging in 1 state (Texas).

SPECIALTY

HARD-TO-FIND ITEMS AND SPECIALTY
STORES. SHOP ON THE GO

LITTLE WORLD GIFTS STORE
The free Little World Gifts iPhone app lets you browse a range of
gorgeous, handcrafted, 3D, interactive digital gifts, and send them
to your friends.
Applicable Devices: iPhone
http://www.littleworldgifts.com
http://www.itunes.com/app/littleworldgifts

MOBILEWEAR
Bluetooth caller ID wristwatch.
View caller ID information and control your cell phone calls from
the face of the watch. This wristwatch is water resistant and lasts
approximately 5-7 days based on conditions of use. The battery
can also be recharged through a USB or universal AC adapter.
http://www.abacuswatches.com

URBAN OUTFITTERS
Receive information on upcoming sales.
WAP (Mobile WEBSITE) http://m.urbanoutfitters.com

KOSHERTOPIA
Find Kosher restaurants near you in New York.
http://www.kosher-ny.com

RESOURCES
TIPS, TRICKS AND RECALLS

PRICE CHOPPER

Get product recalls and advisories by having alerts sent to you via Twitter to your cell phone.

Product Recalls and Advisories.

For information on recalls not listed, go to the USDA Food Safety and Inspection Service.

http://www2.pricechopper.com/recalls

TWITTER http://twitter.com/PC_Recall

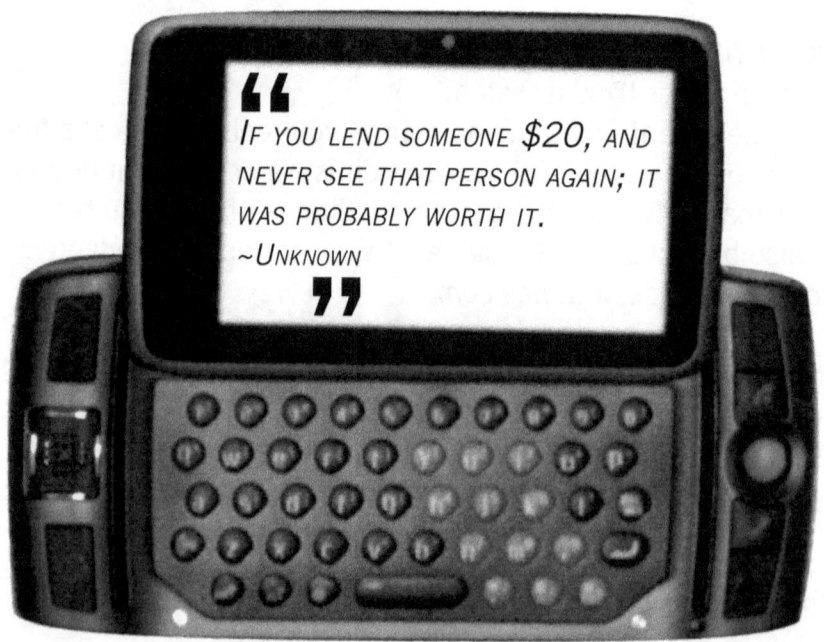

"IF YOU LEND SOMEONE $20, AND NEVER SEE THAT PERSON AGAIN; IT WAS PROBABLY WORTH IT.
~UNKNOWN"

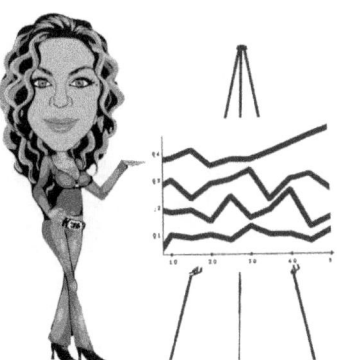

F.U.N.K.Y. STATISTIC:

Cell phone chargers stay warm after you unplug it from the wall because it is actually still draining electricity. According to Future Forests, only 5% of the power drawn by cell phone chargers are actually used to charge phones. The other 95% is wasted when you leave it plugged into the wall, but not into your phone.

"Do not wait for extraordinary circumstances to do good action, try to use ordinary situations."
~Jean Paul Richter

BETTER WORLD

BENEVOLENCE
GIVING TO CHARITIES RIGHT FROM YOUR PHONE

CHARITYCALL
Give to a variety charities.
WAP (Mobile WEBSITE) http://m.charitycall.cc

FREE MILITARY CARE KITS
Military families can send care packages to their loved ones
who are fighting for our country.
Kits include two "America Supports You" large Priority Mail
Flat-Rate boxes, four medium-sized Priority Mail Flat-Rate
boxes, six Priority Mail labels, one roll of Priority Mail tape
and six customs forms with envelopes.
TELEPHONE 1-800-610-8734

iRECYCLE
Donate your cell phone or find over 100,000 recycling and
disposal locations (U.S.) for over 200 materials including
water bottles, cans, motor oil.
Applicable Device: iPhone
http://earth911.com
TELEPHONE 1-800 CLE-ANUP

SALVATION ARMY HAITI
Send money for Haiti relief via SMS.
The SMS Text message donations will go to the Salvation Army's
relief efforts.
TEXT "Haiti" to "45678" (Haiti)
1-800-SAL-ARMY or 1-800-725-2769

uGIVE
Donate to a variety of causes from your cell phone or mobile device.
Applicable Device: iPhone, iPOD Touch
http://www.MobileCause.com
http://iugo.me

CELL PHONES FOR SOLDIERS
Awesome services such as donating money, find a discounted
service, or obtain calling cards for your loved one in the military.
http://www.cellphonesforsoldiers.com
TELEPHONE 1-800-426-1031

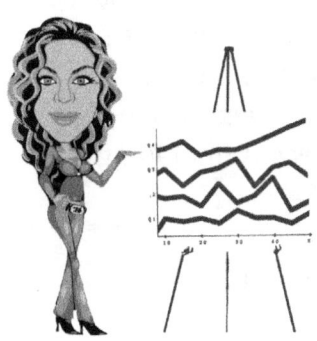

F.U.N.K.Y. STATISTIC:
Rainforests are being cut down at
the rate of 100 acres per minute!

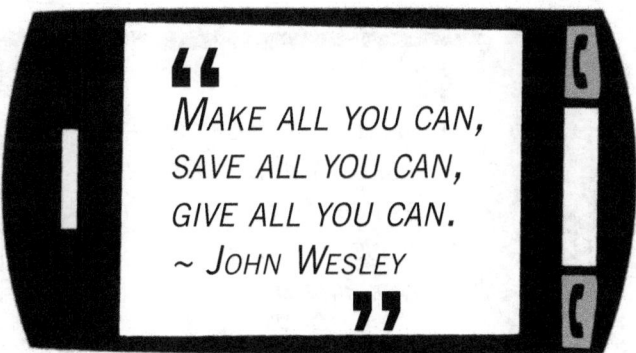

"
MAKE ALL YOU CAN,
SAVE ALL YOU CAN,
GIVE ALL YOU CAN.
~ JOHN WESLEY
"

GOING GREEN

BE EDUCATED AND PROACTIVE TO MAKE OUR PLANET CLEANER AND GREENER

ECOSNOOP APP

Upload photos of environmental mishaps to heighten awareness and spur action.
Applicable Device: iPhone
http://www.ecosnoop.com

iGO GREEN

Green power management solutions for mobile devices. Eliminates standby power not used.
http://www.igo.com

SMARTRECYCLE

Recycle your cell phone, batteries, laptops and any other mobile devices you may have. This is an awesome resource for learning how to become environmentally compliant, what to do with hazardous materials and alternative methods of waste such as obtaining recycling containers to discard waste (including mobile devices).
http://www.batteryrecycling.com
http://www.batteryrecycling.com/SmartRecycle+System
TELEPHONE 1-800-852-8127

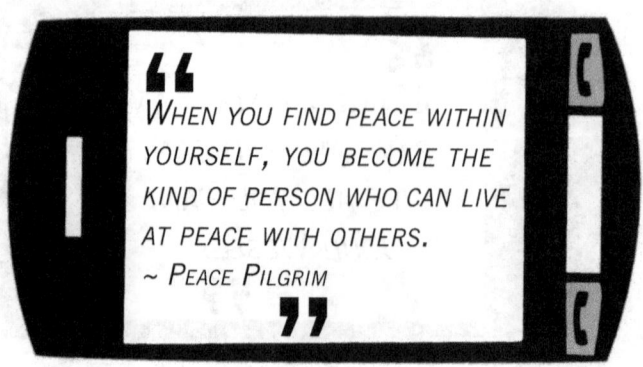

> *WHEN YOU FIND PEACE WITHIN YOURSELF, YOU BECOME THE KIND OF PERSON WHO CAN LIVE AT PEACE WITH OTHERS.*
> *~ PEACE PILGRIM*

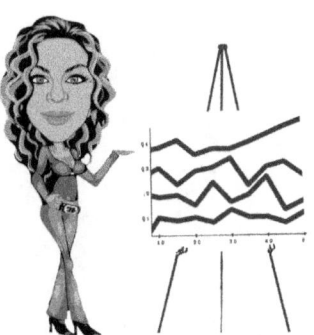

F.U.N.K.Y. STATISTIC:

106,000 aluminum cans are used in the U.S every 30 seconds.

1.14 million paper bags are used in U.S supermarkets every hour.

Recycling one aluminum can saves enough energy to run a TV for three hours -- or the equivalent of a half a gallon of gasoline.

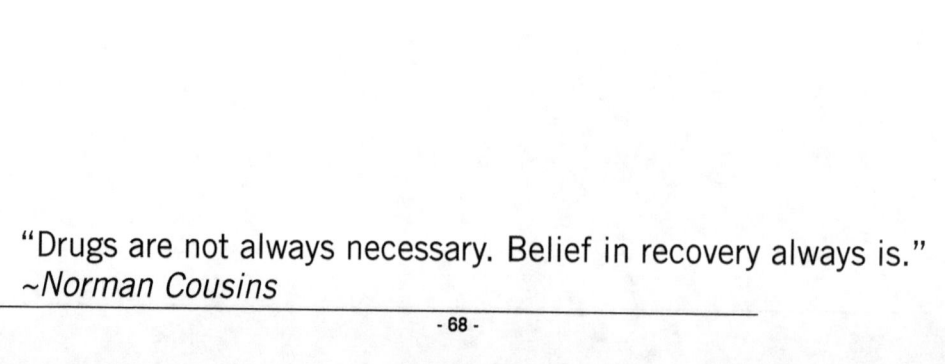

"Drugs are not always necessary. Belief in recovery always is."
~Norman Cousins

WELLNESS

CRISIS

Be empowered to respond immediately to any emergency situation

CENTERS FOR DISEASE CONTROL AND PREVENTION
Receive SMS alerts for Swine Flu, H1N1, and other public health emergencies.
http://www.cdc.gov/mobile
Subscribe: TEXT "HEALTH" to "87000"
WAP (Mobile WEBSITE) http://m.cdc.gov

CODE MOBILE
Missing Persons alert.
Free to sign up at:
http://codeamber.net/codemobile/app/index.php formreg
http://www.codeamber.com/codemobile

FAMILY FIRST ALERTS
Comprehensive emergency service.
http://www.familyfirstalerts.com
For help information, TEXT "HELP" to "88202"

F.U.N.K.Y. TIP:
Everyone knows that the emergency number is 911 in the US, but do you know that the World Wide emergency number is 112? Dial 112 when you are out of your coverage area and it will connect you to an existing network, establishing an emergency number for your location.

HANDS-ONLY CPR
Know what to do if a sudden heart attack occurs.
Applicable Device: iPhone
http://handsonlycpr.org

I AM SAFE
Send "panic" and emergency alerts to multiple contacts.
Your location is recorded for your safety.
Allow a contact to monitor your whereabouts and more for your
personal security and safety.
Applicable Device: iPhone
http://www.iamsafe-mobile.com

NATIONAL ABORTION FEDERATION HOTLINE
Considering abortion and want to talk to someone you can trust?
Get your questions answered here 9:00 a.m. to 7:00 p.m. E.T.,
Monday through Friday; 9:00 a.m. to 3:00 p.m. E.T., Saturday.
TELEPHONE 1-800-772-9100

NATIONAL DOMESTIC VIOLENCE HOTLINE
Experiencing abuse from a teen or adult partner? Not sure if it is
abuse? Call 24 hours a day, 7 days a week.
TELEPHONE 1-800-799-SAFE (7233) or TTY 1-800-787-3224.

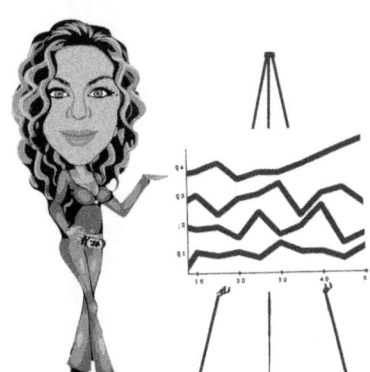

F.U.N.K.Y. STATISTIC:
Anthrax, if you touch it, will infect you, yet inhaling it is even deadlier. Symptoms are like that of a cold or flu, followed by complete respiratory collapse.

NATIONAL SEXUAL ASSAULT HOTLINE
Have you been raped? Not sure exactly what happened?
Call 24 hours a day, 7 days a week.
TELEPHONE 1-800-656-HOPE (4673)

NATIONAL SUICIDE PREVENTION LIFELINE
Call for suicide prevention and emotional crisis.
http://suicidehotlines.com
TEXT "hello crisisline" to "2333"
(Globe) or 211 (Smart)

NATIONWIDE
Exchange automobile accident information, locate an agent,
manage accident photos, assists with obtaining towing
services, process your claims and more.
Applicable Devices: iPhone, iPOD Touch
http://www.nationwide.com/mobile/iPhone-support.jsp
WAP (Mobile WEBSITE) http://www.nationwide.com/mobile/index.jsp
TWITTER http://twitter.com/nationwide
FACEBOOK http://www.facebook.com/nationwide
TELEPHONE 1–877–669–6877

PREGNANCY
Crisis pregnancy support.
TELEPHONE Crisis Hotline 713-HOTLINE▨▨▨
En Espanol 713-526-8088▨▨▨
TeenLine 713-529-TEEN

SEXINFOSF.ORG
Sex education hotline for the youth.
http://sextextsf.org

TELNIC
KFA Technologies created a new .tel domain to help people get near-instant access to emergency services worldwide from any device connected to the internet, including mobile phones.
http://telnic.org
http://sos1.tel

THE NATIONAL RUNAWAY SWITCHBOARD
Are you OR a friend thinking about running away?
Call 24 hours a day, 7 days a week.
TELEPHONE 1-800-RUNAWAY (786-2929)

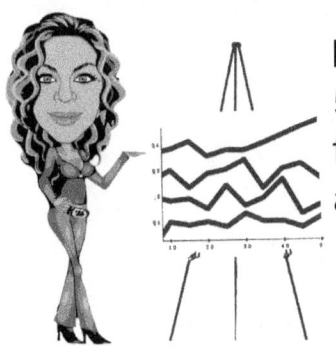

F.U.N.K.Y. STATISTIC:
55% of all deaths caused by firearms in the United States are suicides.

MEDICAL

FIND YOUR RECORDS, HELP CARE FOR A LOVED ONE, OR GET ADVICE & REFERRALS - ALL ON YOUR PHONE

AIRSTRIP
Virtual, real-time remote access and real-time patient information.
http://www.airstriptech.com/Portals/_default/Skins/AirstripSkin/home.aspx

ALLONE MOBILE
Obtain medical records and insurance information for yourself or a senior patient you are caring for anytime.
Applicable Devices: Blackberry, Palm, Nokia, more.
http://www.allonemobile.com

GO SMOKEFREE
Receive motivational SMS text messages to quit smoking.
http://www.gosmokefree.nhs.uk

CALIFORNIA SMOKERS' HELPLINE
Trying to quit? Call for support.
TELEPHONE 1-800-NO BUTTS

EOSHEALTH
Health and wellness programs that can be accessed via mobile phone.
http://www.eoshealth.com/Public/WhoIsItFor/Individuals.aspx
TELEPHONE 1-800-W4L-HELLO (1-800-945-4355)

ESP MOBILE APP
Receive information regarding pain management and clinical support.
Applicable Devices: iPhone, iPOD Touch
http://www.emergingsolutionsinpain.com/index.php?option=com_content&view=article&id=558&Itemid=241

H1N1 INFORMATION
SMS for flu alerts and information.
SMS a TEXT "flu" to "66247"

HOOKUP
California started HookUp "365247", a statewide text-messaging service. The texter can type a ZIP code and get a local clinic referral, as well as weekly health tips.

MD CONSULT
Subscribe to access medical information on your mobile phone's browser.
http://www.mdconsult.com
WAP (Mobile WEBSITE) http://m.mdconsult.com

MEDICA
Cost comparison for a variety of medical care such as chiropractic, clinic, medical equipment, supplies, pharmacy, even same-day surgery.
http://www.mainstreetmedica.com
FACEBOOK http://www.facebook.com/pages/Main-Street-Medica-Mobile

MOBILE COLD & FLU ALERTS
WAP cold and flu alerts.
WAP (Mobile WEBSITE) http://www.4infoalerts.com/wap/robitussin

NATIONAL STD HOTLINE

Think you may be infected with a sexually transmitted disease or HIV? Call 24 hours a day, 7 days a week.

Espanol: 1-800-344-7432 Línea Nacional de las ETS de los CDC El número al que puede marcar es 1-800-344-7432, la cual es una⊠Línea Nacional de las ETS de los CDC. Si se esta preguntando si esta infectada, ¡llámeles ahora! Esta abierto de 8 de la mañana hasta las 2 de la mañana, E.T., los 7 días de la semana. TELEPHONE 1-800-227-8922

OUTBREAKS NEAR ME

Global disease outbreak information.
Applicable Devices: iPhone, Android
http://healthmap.org/outbreaksnearme
http://healthmap.org/en

WEBMD'S MEDSCAPE

Find herbal supplements and generic pharmaceutical brands; search a drug for price comparisons, adverse effects and much much more.
Applicable Devices: iPhone, iPOD Touch
http://www.medscape.com/public/iphone

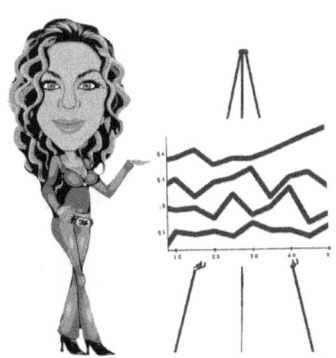

F.U.N.K.Y. STATISTIC:

Daily exercise has been shown to improve blood cholesteral levels, prevent bone loss, boost ones energy level, release tension, improve one's self image.

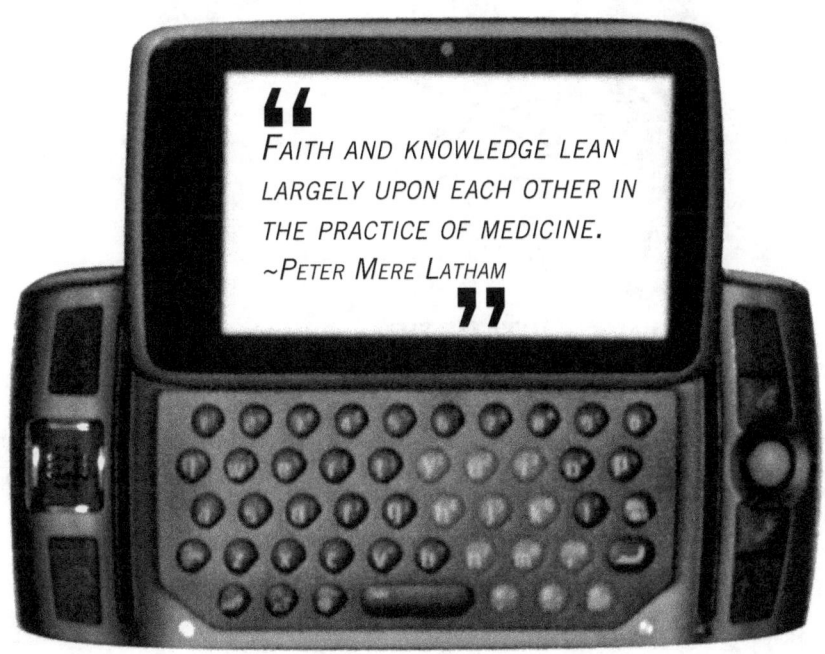

> *FAITH AND KNOWLEDGE LEAN LARGELY UPON EACH OTHER IN THE PRACTICE OF MEDICINE.*
> *~PETER MERE LATHAM*

NUTRITION

WHAT SHOULD YOU EAT AND WHY
ADVICE COMES RIGHT TO YOUR PALM

AETNA STUDENT HEALTH
Count calories and stay fit!
Applicable Devices: Blackberry
http://aetnastudenthealth.com
FACEBOOK http://aetnastudenthealth.com/fb

DIET.COM
Nutrition info sent via SMS.
http://www.diet.com/mobile

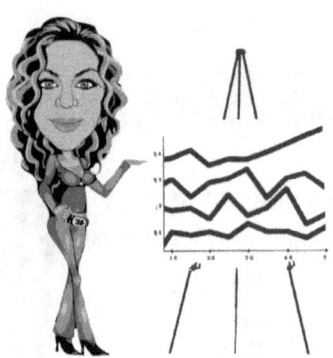

F.U.N.K.Y. STATISTIC:
The crust of a slice of bread has 8 times the amount of cancer fighting antitoxins than the center of the same slice of bread would have.

PRANAYAMA
Breathing techniques to lower your stress.
Applicable Device: iPhone
http://www.saagara.com

THE NATIONAL ALCOHOL AND SUBSTANCE ABUSE INFORMATION CENTER
Think you might have a drinking or substance abuse problem?
Call for support 24 hours a day, 7 days a week.
TELEPHONE 1-800-784-6776

VIRGIN ATLANTIC'S COURSE FOR FLYING WITHOUT FEAR
Receive cheap therapy for fear of flying directly on your iPhone.
Applicable Device: iPhone
http://www.mentalworkout.com

YOGA JOURNAL
15 unique yoga classes to download on your iPhone.
Applicable Device: iPhone
http://www.yogajournal.com/iPractice

SPIRITUAL
FROM BIBLE VERSES TO SPIRITUAL ADVICE

DAILY VERSE
Receive devotional Bible verses every day.
http://www.pray4u.co.uk/daily-bible-text-sms.php

TEXT A BIBLE
Anyone with a cell phone can now text a Bible.

TEXT the word "Bible" to "20222" and wait for confirmation TEXT message.

Send the word "Yes" in response to the confirmation text message.

That's it. You sent a Bible. You'll get a $5 charge on your billing statement.

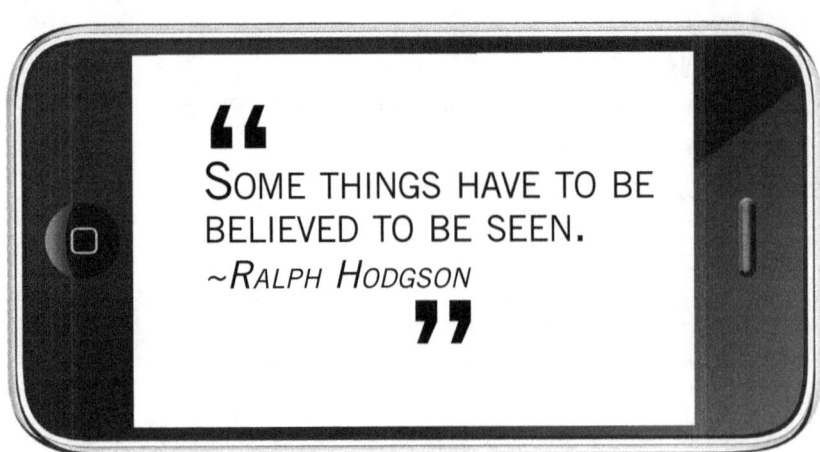

> SOME THINGS HAVE TO BE BELIEVED TO BE SEEN.
> ~RALPH HODGSON

SELF HELP

TAKE CHARGE OF YOUR LIFE WITH STRESS BUSTERS, EXERCISE JOURNALS, HOW-TO MANUALS AND MUCH MORE

GREAT CAREER (FRANKLIN COVEY)
Build your ultimate career by exploring your strengths and talents.
Applicable Device: iPhone
http://www.franklincovey.com/greatcareer

HAPPYTAPPER
Vision board and gratitude journal.
Applicable Devices: iPhone, iPOD Touch
http://happytapper.com

iCONTROL
Interactive home security.
Applicable Devices: iPhone, SMS alerts
http://www.icontrol.com

LIVE HAPPY
Get assistance to stay stress-free in your life.
Applicable Device: iPhone
http://www.livehappyapp.com
TWITTER http://twitter.com/mentalhealtham
FACEBOOK http://www.facebook.com/mentalhealthamerica
TELEPHONE 1-800-273-TALK

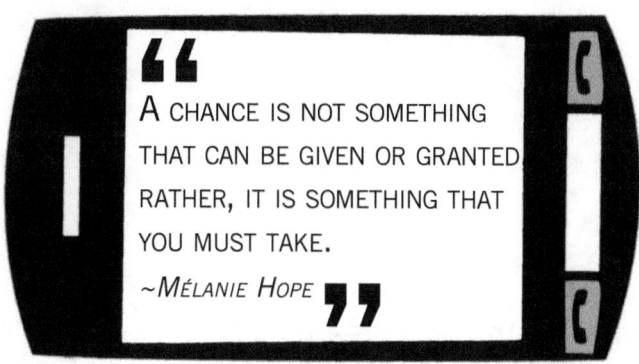

"
A CHANCE IS NOT SOMETHING THAT CAN BE GIVEN OR GRANTED RATHER, IT IS SOMETHING THAT YOU MUST TAKE.
~MÉLANIE HOPE **"**

"Human beings are the only creatures that allow their children to come back home."
~Bill Cosby

PARENTING

RESOURCES
RESOURCES FOR YOUR CHILD OR YOURSELF

MOSIO ANSWERS
Sex Ed via text messaging, and SMS Crime Tips.
http://www.mosio.com/biz/solutions/health
TEXT "ASKMO" your message to "66746"

PEDIDOSER
Easy reference guide for pediatric outpatients.
Applicable Devices: iPhone, iPod, Palm, Android,
Windows Mobile, Blackberry
http://www.pedidoser.com/isilo/buy.html
http://www.pedidoser.com/isilo/index.html

THE BIRDS & BEES TEXT LINE
Text a question to "36263".
Get an answer text back within 24 hours.

THE CENTURY COUNCIL
Reminds teens to make smart decisions about drinking and driving.
http://www.centurycouncil.org

BABY CARE
Baby schedule tracking software for new moms.
Applicable Devices: iPhone, Blackberry
http://www.babblesoft.com/userDetail.php
WAP (Mobile WEBSITE) http://m.babblesoft.com

CHILDHELP NATIONAL CHILD ABUSE HOTLINE
1-800-4-A-Child

TIDE STAIN SOLUTIONS
Fix a stain your child has made. Should you blot or rub, use hot or
cold water. Find immediate solutions.
Applicable Devices: iPhone, Android
http://www.tide.com/en-US/stains/iphonelanding.jspx
http://www.androlib.com/android.application.com-tide-stainbrain-jwti.aspx
FACEBOOK *http://www.facebook.com/Tide*

MONITORING

Monitor your children's web and phone use or track their well-being

MOBILE NANNY PRO
Monitor your children's phone calls, block a number for either SMS or voice, or a website.
Applicable Device: Windows Mobile
http://www.mobile-nanny.com

NET NANNY
Receive SMS alerts when a keyword is identified.
http://www.netnanny.com/mobilE
TWITTER http://twitter.com/NetNanny

PHONE CREEPER
Remotely and secretly monitor and read SMS messages and MORE.
Applicable Device: Windows Mobile
http://forum.xda-developers.com/showpost.
php?p=3977534&postcount=1

TRACE TEXT
Trace the cell phone number or text message back to an owner.
http://www.tracetext.com

F.U.N.K.Y. TIP:
NEVER ignore threats, either verbal or by phone or via text message.

ALERTS

FROM AMBER ALERTS TO PREDATORS TAKE CONTROL OF YOUR CHILD'S SAFETY

AMBER ALERT
Missing child alert.
If you or anyone you know has information or has seen a missing child, call (toll-free 24-hour hotline).
1-800-THE-LOST (800-843-5678)
Applicable Device: iPhone
http://www.zdziarski.com/projects/amberalert
TELEPHONE 1-800-THE-LOST (800-843-5678)

AMBER ALERTS
Wireless Amber Alerts.
http://www.wirelessamberalerts.org
TEXT "AMBER" followed by a space and five-digit ZIP code to AMBER (26237)

F.U.N.K.Y. STATISTIC:
To date, AMBER Alerts have helped to safely recover more than 467 children.

FUN

FUN AND INNOVATIVE WAYS TO TEACH AND ENTERTAIN YOUR CHILD RIGHT ON YOUR PHONE

SANTA TEXT ME
Personalized messages from Santa via SMS.
Donation to the March of Dimes made with every transaction.
http://www.Santatextme.com

F.U.N.K.Y. TIP:
Did you know you can call
1-800-FRUCALL to get a
price check? Follow the simple
instructions to make certain you
are getting a fair price and pay less.

"I have never been in a situation where having money made it worse."
~Clinton Jones

BONUS

TIPS + RESOURCES = BONUS ITEMS

TECHNOBABBLE

By Mélanie Hope

Running ragged seems my lot,

Can't maintain the things I've got.

I can barely keep the pace -

Half my brain's in cyberspace.

The other half is in the sea

Planning out my destiny.

Finding balance seems far-fetched,

With cords and cables all outstretched,

Hanging out in coffee joints,

Going over talking points,

I haven't lost my sanity -

I've found it electronically.

I don't care what others say -

I like my fast-paced life this way.

I like to browse and mail and tweet,

I like to find those deals sweet.

I organize it all with zest,

Doing so much more than all the rest.

It may seem I juggle far too much,

Perhaps I lack that human touch,

But I'm using my technology

To mobilize and set me free.

What once was seen as nerd or geek

Is now considered techno-chic.

BONUS

Crazy gags and goofy tricks that utilize your phone and your imagination

THE REJECTION HOTLINE
Let him down easy by giving this phone number to instead of your own.
TELEPHONE 206-494-0827

KEYS PLEASE (CANADA)
Convince your friends that they can check their blood-alcohol level by breathing into your phone.
OR call 1-877 EZ-ALCO-TEST to breathe into your mobile phone directly and check your blood alcohol level. STAY SAFE!
http://www.keysplease.net
TELEPHONE 1-866-586-KEYS

SPOOFCARD
Spoof your callback number.
http://www.spoofcard.com/?gclid=CIHPxfDkz6ECFQdZbAoddGL5lg

STUPID.COM
Crazy, gag gifts like an old-fashioned handset receiver that can be plugged into your cell phone.
http://www.stupid.com/fun/PLEZ.html

ZEDGE
Send videos of gags, ringtones, and other fun things to your phone.
http://www.zedge.net/videos/170815/a-curve-ball-gag-video

SMS GATEWAY
Reach any cell phone via SMS from any computer.
http://www.clickatell.com/products/sms_gateway.php?cid=13686

IPHONE & IPOD INSURANCE PROTECTION

In the event of any damage or accident be protected by insurance.
Applicable Devices: iPhone, iPOD Touch
http://www.missionrepair.com
TELEPHONE 1-866-638-8402

ARE YOU INTERESTED

Dating app.
Applicable Device: iPhone
*http://itunes.apple.com/WebObjects/MZStore.woa/wa/
viewSoftware?id=307930478*

DATECHECK

Dating app to check for criminal background. Even find out
information about birthdates, relatives, residence.
Applicable Devices: Android, iPhone
http://www.intelius.com/mobile

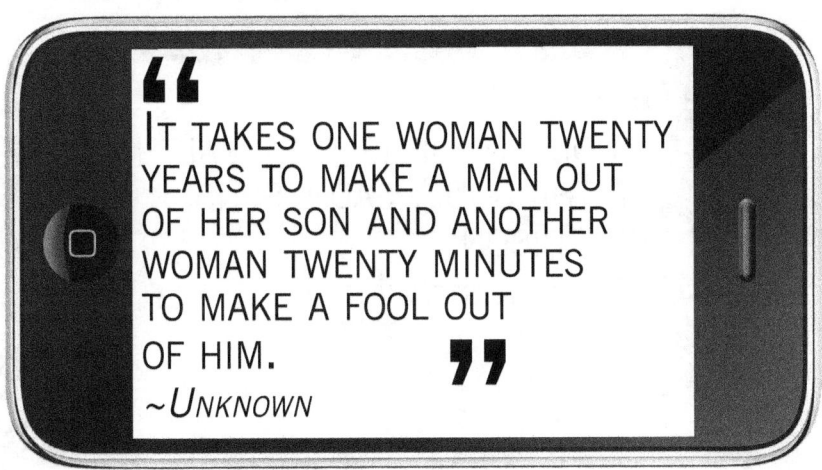

" IT TAKES ONE WOMAN TWENTY YEARS TO MAKE A MAN OUT OF HER SON AND ANOTHER WOMAN TWENTY MINUTES TO MAKE A FOOL OUT OF HIM. **"**
~UNKNOWN

NEARU
Get money-saving coupons sent to your mobile phone.
http://www.nearusearch.com
TEXT "HELP" to "63278" (NearU)

CASH OLD PHONE
Trade your old cell phone for cash.
Applicable Device: iPhone, Treo, Blackberry and More
http://www.casholdphone.com

DOG PARK FINDER
Find a dog park near you.
Applicable Device: iPhone
http://www.dogparkusa.com/iphone/dog-park-finder

MONEYGRAM
Find nearest Money Gram near you.
Applicable Device: iPhone
http://www.moneygram.com/MGICorp/campaigns/mobileapp

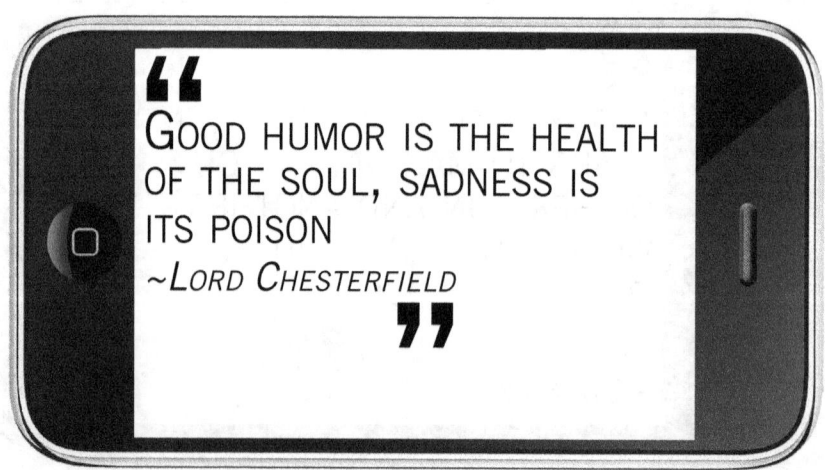

> **G**OOD HUMOR IS THE HEALTH OF THE SOUL, SADNESS IS ITS POISON
> ~*LORD CHESTERFIELD*

F.U.N.K.Y. FIRE SAFETY TIPS:

Fire safety tips featured in "Pella's Close the Door on Fire!" program are:

1. Have a fire safety plan written where EVERYONE can see it in your office or home. Review it with everyone!

2. Identify your exits! Then have them highlighted in your Fire Safety Plan.

3. Use your smoke alarms and make certain they work before a fire acutally happens.

4. Make certain your exits entries are free from any clutter!

5. Have emergency numbers listed and posted in your Fire Safety Plan! In the event of an emergency.

6. When escaping during a fire remember to crawl and stay low on your hands and knees.

7. Review the plan with EVERYONE that would act in the event of a fire.

Fire Safety iPhone App.
Review details at:
http://www.safetymedia.co.uk/acatalog/Fire_Safety_iPhone_App.html

PART 3

ON-THE-GO FREEDOM: MOBILE PRODUCTIVITY

- **BUSINESS**
  Productivity
  Shipping/Tracking
  Job Search
  Networking

- **FINANCE**
  Banking
  Budgeting
  Insurance

- **TRAVEL & NAVIGATION**
  Connectivity
  GPS
  Maps
  Roadside
  Assistance
  Flight/Hotel
  Airlines

  Transportation
  Insurance
  Alerts
  Safety
  Tips

- **EDUCATION**
  Resources
  Alerts
  Directories
  Web

- **REFERENCE**
  Manuals &
  Instruction
  Food
  Law
  Animals

"Productivity is never an accident."
~Paul J. Meyer

BUSINESS

PRODUCTIVITY

Tips and apps to increase your productivity on the go

ALERTS.COM
Customized SMS alerts for Craigslist, birthday reminders, school alerts, wake up calls, movie alerts, job alerts, flights and much more! Also, alerts for oil change and discounts to the places such as Jiffy Lube.
http://www.alerts.com

CALLIFLOWER
Visually manage your conference calls, see who is online, or record a call (adding any notes you may have taken).
Applicable Device: iPhone
http://calliflower.com

DAYLITE TOUCH
Business productivity manager. Keep up with your projects, contacts, sales and schedule in your pocket.
Applicable Devices: iPhone, iPod Touch, Blackberry
http://marketcircle.com

EPLIXO
Video conference.
Install directly from the mobile site.
Applicable Device: Windows Mobile
WAP (Mobile WEBSITE) http://eplixo.com/m

QUICK OFFICE
View, edit and create Office documents on your device.
Applicable Devices: Blackberry, iPhone, and Palm
http://www.quickoffice.com
TWITTER - http://twitter.com/quickoffice
FACEBOOK - http://www.facebook.com/Quickoffice

EVERNOTE
Note taking app that works via voice, email, web page clip, image and more!
Applicable Device: Android
http://www.evernote.com

FMTOUCH
Deploy relational databases.
Applicable Devices: iPhone, Blackberry
http://www.filemaker.com/products/iphone/index.html
http://developer.filemaker.com/solutions/detail/?item=solution.1000 0002753&sol_region_amr=1

MICROSOFT ONEAPP
New software application that enables feature phones to access mobile apps like Facebook, Twitter, Windows Live Messenger, and other popular apps and games.
http://www.microsoft.com/oneapp
App Gallery:
http://www.microsoft.com/oneapp/product/appgallery.aspx

NTRCONNECT VIEWER
Remote access to your PC or MAC.
Applicable Devices: iPhone, Windows Mobile
http://www.ntrconnect.com
https://na.ntrconnect.com/web/mobilityasp?r=0.3665903434310 952&redir=0

MOBILE CRM EXPRESS
Awesome mobile customized Customer Relationship Managment (CRM) solution! For those of you who need access to a flexible yet highly reliable and functional CRM tool - this is it!
http://www.akvelon.com

PRINTERON

Print your web email from Hotmail®, Yahoo!® Mail and Google Mail™ from your mobile phone. This includes attachments.
Applicable Devices: BlackBerry, iPhone, (Pocket PC) Windows Mobile
http://www.printeron.net

SELFORGANIZER.COM

SMS calendar reminders.
http://www.selforganizer.com

SYMANTEC™ DEEPSIGHT™ ALERT SERVICES

SMS Alerts to stay up-to-date about computer viruses using this alert service. Alerts for vulnerabilities, malicious code, spyware, adware and what to do.
https://alerts.symantec.com/Default.aspx

TOODLEDO

Online to do list! Can import from Palm, Microsoft Outlook, and Apple iCal.
http://www.toodledo.com
http://www.toodledo.com/info/iphone.php

ZOHO MOBILE

Powerful business applications such as Microsoft Office suite word processor, spreadsheet, Wikis, email, invoice.
Applicable Devices: iPhone, Android, Blackberry, Windows Mobile, Nokia
WAP (Mobile WEBSITE) https://mobile.zoho.com/login
TWITTER http://twitter.com/zoho

SHIPPING & TRACKING
FOLLOW YOUR PACKAGES FROM ANYWHERE

TEXTRACK
UPS or FedEx tracking.
http://www.textrack.com/cgi-bin/home.cgi

TRACKMYSHIPMENTS
Track a shipment from Amazon, DHL, FedEx, UPS, USPS
and others.
Applicable Devices: SMS or iPhone
http://www.trackmyshipments.com

TRACKTHIS
Track a package for FedEx, USPS, DHL, or UPS using email,
SMS or Twitter.
http://www.usetrackthis.com

UPS MOBILE
Manage and track your shipments.
Applicable Device: iPhone
http://www.ups.com/iphone/?gclid=CPf8343L6Z4CFShGagodfRYOKg

UPS TRACKING ALERTS
Track or ship a package and have SMS alerts sent to your
cell phone.
Per UPS - follow these steps:
1. Select the message option on your mobile phone.
2. Select write message.
3. Input your UPS tracking number. a. If you insert "UPS" before
   the tracking number you will automatically be notified within
   minutes on delivery. b. If you omit "UPS," you are informed of
   the latest status of your shipment either in transit or with full
   delivery details.
4. Press send.
5. Enter UPS phone number
   http://www.ups.com/content/gb/en/tracking/tracking/sms/index.
   html?WT.svl=SubNav

JOB SEARCH

FIND, APPLY FOR, AND RESPOND TO JOB OPPORTUNITIES ALL FROM YOUR MOBILE DEVICE

LINKEDIN MOBILE

Update your professional profile for potential employers to view.
Applicable Devices: iPhone, Blackberry, Palm
http://www.linkedin.com/static?key=mobile
WAP (Mobile WEBSITE) http://m.linkedin.com

LINKUP

Mobile job search engine.
Applicable Devices: iPhone, Android
http://www.linkup.com/mobile

MONSTER MOBILE

Search and apply for jobs on the go!
http://promotions.monster.com/mobile

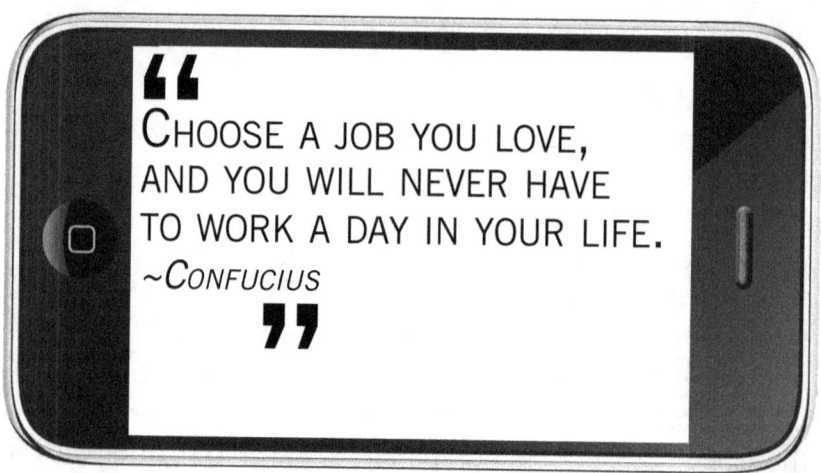

"
CHOOSE A JOB YOU LOVE, AND YOU WILL NEVER HAVE TO WORK A DAY IN YOUR LIFE.
~CONFUCIUS
"

NETWORKING

MOBILE BUSINESS CARDS AND BUSINESS NETWORKING APPS

DUB

Mobile Business Card.
Applicable Devices: iPhone, Blackberry, Android
https://www.dubmenow.com/downloadnow
WAP (Mobile WEBSITE) http://m.dubmenow.com

iONCE (PAGEONCE)

Alerts for email, finance, social networking, utilities, travel and shopping; changes in flight status, itineraries, accounts, cell minutes. Manage numerous accounts and passwords centrally. One cool feature is the ability to "nuke" all your personal information if your device is lost or stolen.
Applicable Devices: iPhone, iPOD Touch
http://www.pageonce.com

SEND VCARD VIA SMS

Send a business card via SMS or wirelessly.
http://*www.ipipi.com/help/send-vcard-using-sms.htm*

WORLDCARD MOBILE

Business card reader app.
http://worldcard.penpowerinc.com

"The golden rule for every business man is this: Put yourself in your customer's place."
~Orison Swett Marden

FINANCE

BANKING

ATM HUNTER

MasterCard sends text messages ATM locations to cardholder's cell phone.
Applicable Device: iPhone
http://www.mastercard.com/us/personal/en/cardholderservices/mobileservices.html
TELEPHONE 1-877-FIND-ATM

MERCHANTWARE MOBILE

Process credit card transactions from your mobile device.
Applicable Devices: Blackberry, iPhone
http://merchantwarehouse.com
http://merchantwarehouse.com/download.html

PAYMATE

PayMate powers your phone to instantly send and receive money, pay for retail purchases, monthly utility bills, flight & movie tickets with unmatched ease, speed and safety.
http://www.paymate.co.in/web/index.aspx

PAYPAL'S SEND MONEY APP

Send money via your PayPal account.
Examples: Send 10.99 to 2125551981 10.99 21255519815 name@domain.com Send 5 to name@domain.com and much more.
Applicable Devices:iPhone, Android, Blackberry
https://www.paypal.com/mobile
SMS Keywords from PayPal: https://www.paypal.com/cgi-bin/webscr?cmd=xpt/Marketing/mobile/MobileAdvancedFeatures-outside
Check your PayPal balance - TEXT "bal" or balance to "729725"
Send money Send a text to "729725" (PAYPAL). Specify the amount and the recipient's phone number or email address

PRECASH
Prepaid debit card.
Applicable Devices: SMS Text Alerts
http://www.visionprepaid.com
http://www.precash.com

SYBASE 365
Mobile banking solution.
Applicable Device: iPhone
http://www.sybase.com/mobileservices
http://www.sybase.com/365

USAA MOBILE APP
Access your insurance, banking and investment accounts.
Applicable Device: iPhone
https://www.usaa.com/inet/ent_utils/McStaticPages?key=
usaa_mobile_main

VIRTUAL WALLET
Online banking for college students that sends SMS alerts for possible overdrafts, spending tracker, upcoming bill calendar reminders, alerts to parents when account is at or below a specified threshold. Allows parents to transfer money to their child's account, enable overdraft protection and receive digital receipts.
Phone Devices: iPhone Or SMS
*https://www.pnc.com/webapp/unsec/Blank.do?siteArea=/PNC/Pncbk/
Virtual+Wallet+Student*
TWITTER www.twitter.com/pncvwallet

VISA EUROPE
Receive mobile alerts for your credit card activity - especially when traveling.
http://www.visaeurope.com

BUDGETING

APPS TIPS AND TRICKS TO KEEP YOUR MONEY UNDER CONTROL

YODLEE MOBILE SMS

A solution to provide budgeting tools, current value estimates of real estate holdings, mobile and email alerts and other value added features.

With simplified account switching, next generation funds transfer, and mobile access, the Yodlee Payments Suite helps financial institutions save money and make money via the online channel. Yodlee's SMS alerts help consumers avoid overdraft fees by alerting them, in real-time, when their balance is close to falling below zero. Yodlee uniquely leverages the Internet and a patented technology to collect, consolidate, and present real-time financial data from virtually any financial source, including bank accounts, brokerages, credit cards, bills, rewards programs, and more.
http://www.yodlee.com

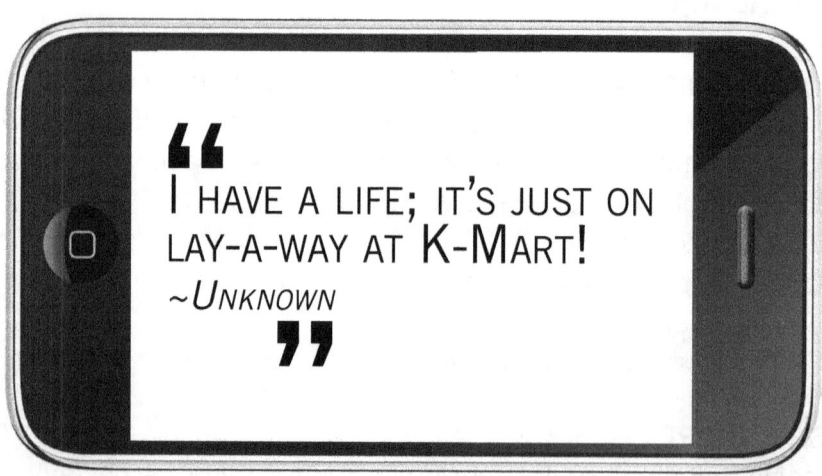

"I HAVE A LIFE; IT'S JUST ON LAY-A-WAY AT K-MART!
~UNKNOWN"

INSURANCE

FROM TRAVEL HEALTH AND SAFETY TO LIFE INSURANCE, YOUR QUESTIONS ARE ANSWERED ON YOUR PHONE

ALLSTATE
Insure your mobile device or cell phone.
http://www.allstate.com
TELEPHONE 1-800 Allstate

iPHONE OR iPOD INSURANCE
Full service iPhone or iPod repair insurance.
Applicable Devices: iPhone, iPOD
http://www.missionrepair.com
TELEPHONE 1-866-638-8402

MOBILE DEFENSE
Real-time tracking & remote locking if you lose your mobile phone.
Applicable Device: Android
https://www.mobiledefense.com
https://www.mobiledefense.com/download

MY WEBWILL
Online service that allows you to make decisions about your Internet accounts in the event of your death.
https://www.mywebwill.com
https://www.mywebwill.se
TWITTER http://twitter.com/mywebwill
FACEBOOK http://www.facebook.com/pages/My-Webwill

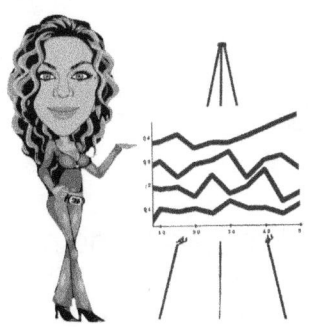

F.U.N.K.Y. STATISTIC:
Motor oil never wears out, it just gets dirty. Oil can be recycled, re-refined and used again, reducing our reliance on imported oil.

"The world is a book, and those who do not travel read only one page."
~St. Augustine of Hippo

TRAVEL
& NAVIGATION

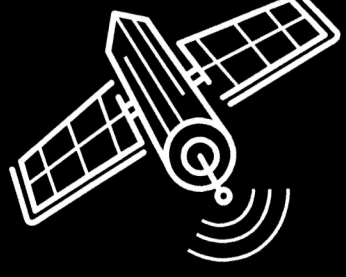

CONNECTIVITY

Always know where the best signals are located

BOINGO
Over 125,000 WiFi Hotspots.
Applicable Devices: Android, Blackberry, iPhone,
Windows Mobile, Nokia
http://mobile.boingo.com
http://mobile.boingo.com/download

GOGO INFLIGHT
Get Internet access on selected airlines while in flight.
Applicable Devices: Blackberry, Windows Mobile, iPod Touch,
iPhone, Palm, Nokia
http://www.gogoinflight.com

POINTABOUT
Mobilize your RSS, XML feed, and HTML content.
Applicable Devices: iPhone, Android
http://www.pointabout.com

SAMSUNG MONDI
WiMAX-enabled handheld device (Mobile Internet Device).
Buy from Clearwire, Best Buy, or www.Samsung.com.

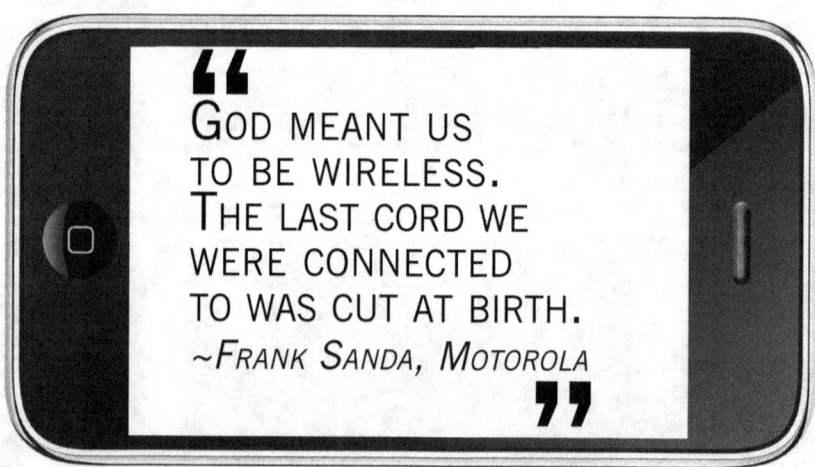

"
God meant us
to be wireless.
The last cord we
were connected
to was cut at birth.
~Frank Sanda, Motorola
"

GPS

ALWAYS KNOW WHERE YOU ARE OR WHERE YOU ARE GOING

GPS TRACKING
GPS Tracking apps.
Applicable Devices: iPhone, Android
http://www.locimobile.com

iEXIT
Find the nearest freeway exit, a gas station, hotel, grocery store, auto repair shop and more - even if you have no cellular signal.
Applicable Device: iPhone
http://www.iexitapp.com

LIFEINPOCKET
GPS, navigation, location message, sync capability and much more to your phone free of any charge. Free text messaging to and from a PC. A very cool voice-guided navigation makes it hands-free, eyes-free and safe. Features include GPS, navigation, actionable location messaging, friends/family locator, local listing/reviews, satellite map, street view, cheap gas finder, news, video, traffic, Twitter, emails, IMs, Facebook, MySpace, LinkedIn, YouTube, music, movies, local events, TV guide, weather, finance, banking, sports, flights, magazines, fashion, cooking, transportation, travel, yellow/white pages, dictionary, translation, Bible and much more real-time information and services.
Applicable Device: Blackberry
http://lifeinpocket.com
WAP (Mobile WEBSITE) http://lifeinpocket.com/s.html

TRACKPRO

Vehicle tracking system.
Applicable Device: iPhone
http://www.autocoptrackpro.com
WAP (Mobile WEBSITE) http://www.autocoptrackpro.com/vts/
default.aspx

WAZE - WAY TO GO

Waze is a free, turn-by-turn GPS application that uses crowd sourcing to detect traffic conditions in real-time. A social network for motorists.
Applicable Devices: iPhone, Android, Windows Mobile
http://www.waze.com/download

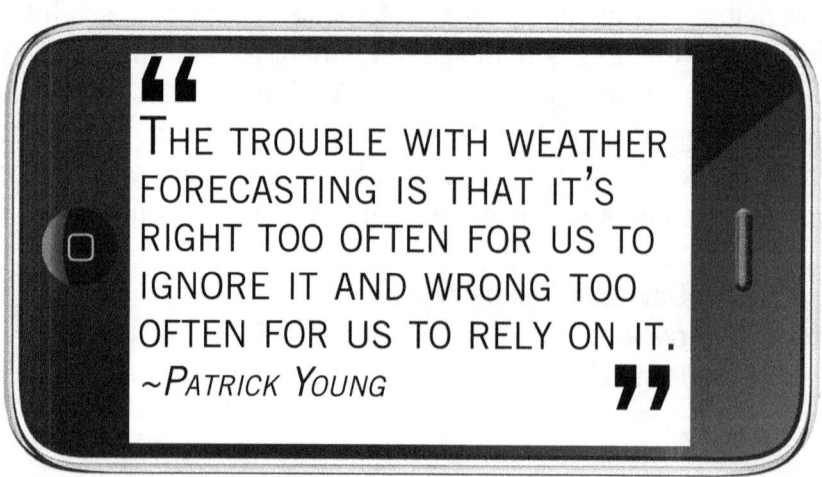

> **"**
> THE TROUBLE WITH WEATHER FORECASTING IS THAT IT'S RIGHT TOO OFTEN FOR US TO IGNORE IT AND WRONG TOO OFTEN FOR US TO RELY ON IT.
> ~PATRICK YOUNG **"**

MAPS

EASY MAPS AND LOCATORS WHEN YOU NEED THEM THE MOST

NAVIGON
Awesome app that provides a 3D real view of highways, interchanges and road sign information. Text-to-speech gives driving directions and street names (even in multiple languages). Other features include lane change assistance, speed warnings, & traffic information!
Applicable Device: iPhone
http://www.navigon.com

WIKITUDE
Augmented reality travel guide.
Applicable Devices: iPhone, Android, Palm
http://www.mobilizy.com

INRIX TRAFFIC
Real-time traffic conditions.
Applicable Devices: iPhone, Android
http://www.inrixtraffic.com
TWITTER http://twitter.com/INRIX

PHAROS GPS
GPS Solutions AND now even mobile cell phones.
Applicable Device: Windows Mobile
http://www.pharosgps.com

ROADSIDE ASSISTANCE

Use your cell phone to gain roadside assistance

511 FREEWAY AID
This Bay Area service allows motorists on the freeway to call from their cell phone for roadside aid.
http://511.org/freewayaid.asp

IGAAUGE
AAMCO Auto Service.
Applicable Device: iPhone
http://www.aamco.com/iGaauge_main.asp

GEICO GLOVEBOX
Insurance, roadside assistance and more.
Applicable Device: iPhone
http://appshopper.com/utilities/geico-glovebox

F.U.N.K.Y. TIP:
Vessel Identification System: Make sure to register your vehicle and that it is legally identifiable.

FLIGHT & HOTEL

Track your flight status and book your hotel room on the run

EXPEDIA'S TRIP ASSISTANT
Receive flight and hotel info directly on your cell as SMS OR download mobile iPhone app in iTunes store.
http://www.expedia.com/daily/promos/tools/iphone/default.asp

HOTELZON HOTEL
Hotelzon, based in Finland, has wholly owned operations in Sweden, the United Kingdom and China.
Applicable Devices: BlackBerry, iPhone
http://www.hotelzon.com/mobile

HYATT HOTELS AND RESORTS
Book a night, make reservations.
http://hyatt.com
http://mobile.usablenet.com/mt/hyatt.com/hyatt/index.jsp

TRIPIT
Forward your travel confirmation emails to anyone needing your itinerary - even yourself!
http://www.tripit.com
TWITTER - http://twitter.com/tripit

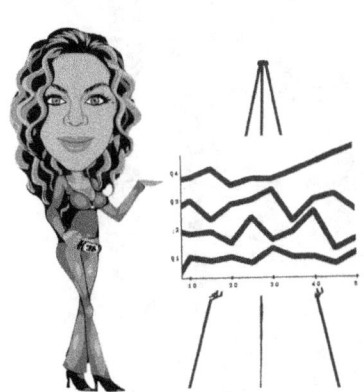

F.U.N.K.Y. STATISTIC:
Average number of people in an airborne vehical over the U.S at any given hour: 61,000.

OUTSIDE A HOTEL:
HELP! WE NEED
INN-EXPERIENCED PEOPLE
~UNKNOWN

AIRLINES

RESEARCH AND RESERVE YOUR FLIGHT RIGHT
FROM YOUR PALM

AMERICAN AIRLINES (ALSO IN SPANISH)
Book a flight and receive your boarding pass on your mobile device.
See http://www.aa.com/mobileboarding for details.
WAP (Mobile WEBSITE) http://www.aa.com/mobile

UNITED AIRLINES
Mobile boarding passes are now in pilot phase.
http://mobile.united.com

US AIRWAYS FLIGHT INFORMATION
Enroll in dividend miles to receive flight information.
http://www.usairways.com

TRANSPORTATION
RESERVE A RIDE ANYWHERE ANY TIME

AVEGO
Green Carpool app that allows drivers and ride seekers to find each other using GPS.
Applicable Device: iPhone
http://www.avego.com

DESIGNATED DRIVER SERVICES - iPHONE APP
Taxi and other transportation services for those times you shouldn't drive yourself.
Applicable Device: iPhone
http://www.designateddriverapp.com

DESIGNATED DRIVER SERVICES - LOS ANGELES
Los Angeles greater area designated driver services.
TELEPHONE 1-800-670-0959

iSIXT
Book a rental car directly from your mobile phone when travelling in Europe.
INSTRUCTIONS:
1. Open Market on your Smartphone
2. Search for "Sixt"
3. Download and install the Sixt application
Applicable Devices: Android, Blackberry, iPhone
http://dk.sixt.com/php/res?language=en
WAP (Mobile WEBSITE) iPhone- iphone.sixt.com; Blackberry - blackberry.sixt.com

ZIPCAR
Find a Zip Car near you!
Applicable Device: iPhone
http://www.zipcar.com

TAXI MAGIC
Book a taxi from your cell phone.
Applicable devices: iPhone, Blackberry, Palm, And Android
http://www.taximagic.com

RIDE CHARGE
Book a taxi, shuttle, and more.
http://www.ridecharge.com

F.U.N.K.Y. TIP:
Letting a car idle wastes fuel.
Rolling down your windows is
more fuel efficient than using
the air conditioner.

INSURANCE

FROM TRAVEL HEALTH AND SAFETY TO LIFE INSURANCE, YOUR QUESTIONS ARE ANSWERED ON YOUR PHONE

mPASSPORT

Receive health and safety tips while traveling.
Video demo at http://www.youtube.com/watch?v=4TlCTdJqwYs
http://www.mpassport.com
http://www.mpassport.com/support.cfm

HTH TRAVEL

Access travel medical insurance from your mobile phone when traveling abroad. Currently there are 6,000 hospitals and doctors in 180 countries that are accessible via HTH Worldwide providing health insurance.
Applicable Devices: iPhone or mobile web
http://www.hthtravelinsurance.com
TELEPHONE 1-888-243-2358 or Outside U.S. 610-254-8700
WAP (Mobile WEBSITE) http://www.mpassport.com/mobile/index.cfm

ALERTS
KEEP YOUR FAMILY INFORMED OF YOUR WHEREABOUTS

ARRIVEDOK
Inform your family, friends or colleagues if you have landed on time at a destination airport via SMS or email alerts.
http://www.arrivedok.mobi/hello

CAVU DESCRIPTION
Mobile surveillance solution for IT professionals.
Applicable Devices: iPhone, iPOD Touch
http://www.cavu.me/buy

ENROUTEHQ
Share your travel or tip itinerary and info with family and friends.
Applicable Device: iPhone
http://www.enroutehq.com

MICRO VBB
Vehicle security to locate your car, intrusion alerts, theft alerts and more.
http://www.microvbb.net

SUPER STOP USA
Sign Up Now for Gas Alerts when prices goes up by about 20%.
TEXT "GAS" to "25827"

ORBITZTLC ALERTS
If you use Orbitz for your travel arrangements, you can also receive alerts and notify up to six contacts via SMS.
http://www.orbitzandgo.com/tlc

TRAPSTER
Tells you where speed traps are located.
Applicable Devices: iPhone, Android, Nokia, Windows Mobile, Blackberry, Palm
http://www.trapster.com

TRIPCASE
Travel app to keep those you desire abreast of your flights and much more!
Applicable Devices: iPhone, Blackberry, Windows Mobile
http://www.tripcase.com

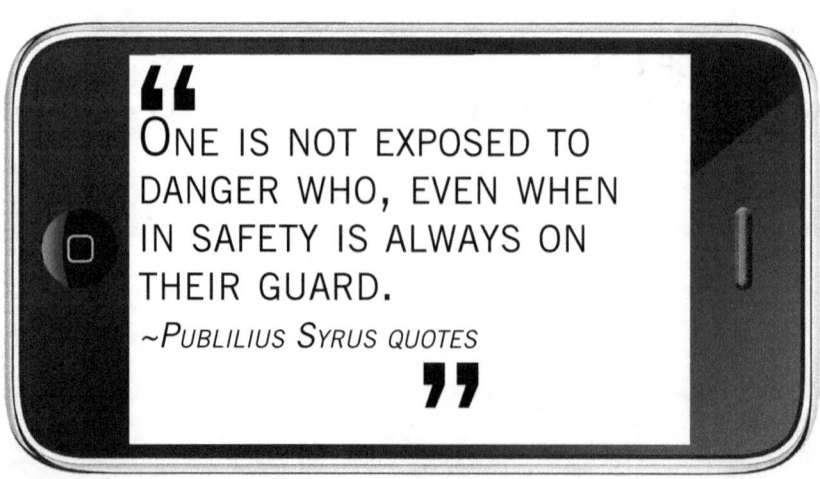

> **"**
> ONE IS NOT EXPOSED TO DANGER WHO, EVEN WHEN IN SAFETY IS ALWAYS ON THEIR GUARD.
> ~PUBLILIUS SYRUS QUOTES
> **"**

SAFETY
Aᴅʜᴇʀᴇ ᴛᴏ ᴄᴇʟʟ ᴘʜᴏɴᴇ ʟᴀᴡs ᴀɴᴅ ᴀʀʀɪᴠᴇ ʜᴏᴍᴇ sᴀғᴇʟʏ

ZOOMSAFER
This mobile software provides hands-free services such as gauging your driving speeds (and disables your cell phone until your car is stopped), auto replies to SMS messages, answers email and phone calls, announces caller ID by voice for greater safety!
Applicable Device: Blackberry
TWITTER http://twitter.com/Idrivefocused
FACEBOOK http://www.facebook.com/pagesZoomSafer/102906602 325?ref=ts

DRIVESAFE.LY
Reads SMS Text messages and emails out loud so you don't have fiddle with your cell phone while on the road.
Applicable Devices: Blackberry, Android, Windows Mobile, And iPhone
http://www.drivesafe.ly/download

TIPS

THE INS AND OUTS OF TRAVEL AND PLANNING ALL AT YOUR FINGERTIPS

POWER TIPS YOU CAN USE!
Convenient tips whether you are in your home city or traveling.

Google Help: TEXT the word "help" to "46645" (Google) and receive simple
instructions on how to use Google's SMS
based info service.

Movies: TEXT the word "m <<your zip code>>" to "46645" (Google) you'll get movie showtimes playing in your area.

Weather: TEXT the word "W <<your zip code>>" to "46645" (Google) will get you the weather forecast for your area.

Pizza: TEXT the word "Pizza"
<<your zip code>>" to "46645" (Google) to get Pizza listings in your area.

Calculator: TEXT the word "3*5" to "46645" (Google) to get calculations, and this example - "3" multiplied by "5" to get the result of "15". Get the idea?

Conversions: TEXT the word "20 miles in kilometers" to "46645" (Google) and you will receive a conversion of "32.18688 kilometers".

Translation: If you find you need translation assistance text the word "T" (for "translate") to "46645" (Google).
Example: "t how are you from English to German" and see the results "Translation: 'how are you' in English means 'wie geht es dir' in German."

"Study without thought is vain; thought without study is dangerous."
~ *Confucious*

EDUCATION

RESOURCES
SCHOOL PLANNING, HOMEWORK HELP AND EDUCATIONAL WEBSITES

AARDVARK
Ask any question, get an answer.
Applicable Device: iPhone
http://www.itunes.com/appstore
http://vark.com/iphone

CAR TALK
Have a question about your car?
Applicable Devices: iPhone
http://www.cartalk.com/content/timekill/car-talk-iphone-app.html

DATATEL MOBILE
See a list of courses and descriptions, professor information, roster, class assignments, events, directory, and much more. Find your buildings or campus.
Students, faculty, and alumni Integrate with Facebook, LinkedIn, IM, and more from phone to phone.
Applicable Devices: iPhone, iPod, Blackberry, Android, Windows Mobile, Nokia, Palm
http://www.datatel.com

TEXT A LIBRARIAN
Ask a librarian a question directly from your cell phone and receive an answer.
http://www.textalibrarian.com
TWITTER - http://twitter.com/textalibrarian
FACEBOOK - http://www.facebook.com/textalibrarian

HEADLIGHT
If you need a headlight replaced but have no manual on hand, use Headlight to find instructions.
Applicable Device: iPhone
http://www.sylvania.com

ALERTS
Is your child's school closed?

E2CAMPUS
SMS emergency notification system for all schools - K-12 through college.
http://www.e2campus.com

e2CAMPUS EMERGENCY NOTIFICATIONS
Over 600 schools and universities use this system for to send emergency alerts via SMS, Facebook, Twitter, email and much more!
http://www.e2campus.com

REDSKY E911 SOLUTIONS
Subscribe to receive university, school or employer police emergency alerts or building closure notices.
http://www.redskye911.com

DIRECTORIES
Find a phone number or directions fast

1-800-FREE411
Call for free 411 and even driving directions.
http://www.free411.com
TELEPHONE 1-800-FREE411

411.COM
Search for 411 or map directions information from your cell phone; also has a reverse lookup.
http://www.411.com

DEXKNOWS
Seniors who are not able to get to a location conveniently can dial or use the mobile app (even SMS) to find a service or referral. 1800calldex is currently available for searches in AZ, CO, IA, ID, MN, MT, ND, NM, OR, SD, NE, UT, WA & WY.
Applicable Devices: iPhone, Blackberry
http://www.dexknows.com/mobile
http://www.dexknows.com/business_profiles/village_senior_care-b218635
TEXT "DexNow" (339669) with your local business search and get quickly!
http://m.DexKnows.com

REFPRIVUS MOBILE
Free caller name lookup upgrade for Windows Mobile app Enhanced Caller ID services.
Applicable Devices: Windows Mobile, iPhone, Android, Blackberry, Palm
http://www.privusmobile.com
http://www.privus.mobi

WEB
SEARCH ENGINES ON THE GO

BING
Web search engine.
Includes voice search capabilities which have already found their way onto Windows Mobile and BlackBerry devices.
Applicable Device: iPhone
http://www.discoverbing.com/mobile/devices/iphone.html

QMOBILESEARCH
All-in-one search engine.
Applicable Devices: iPhone
http://iqmobilesearch.com

PLUCKER
Read websites when you are not connected to the Internet or can't find WiFi access.
Applicable Devices: iPhone, Palm, And Windows Mobile
http://www.plkr.org

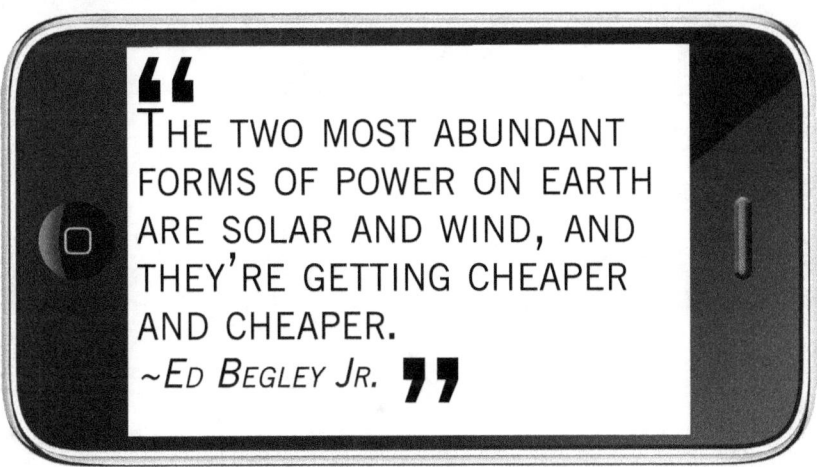

"
THE TWO MOST ABUNDANT FORMS OF POWER ON EARTH ARE SOLAR AND WIND, AND THEY'RE GETTING CHEAPER AND CHEAPER.
~ED BEGLEY JR. "

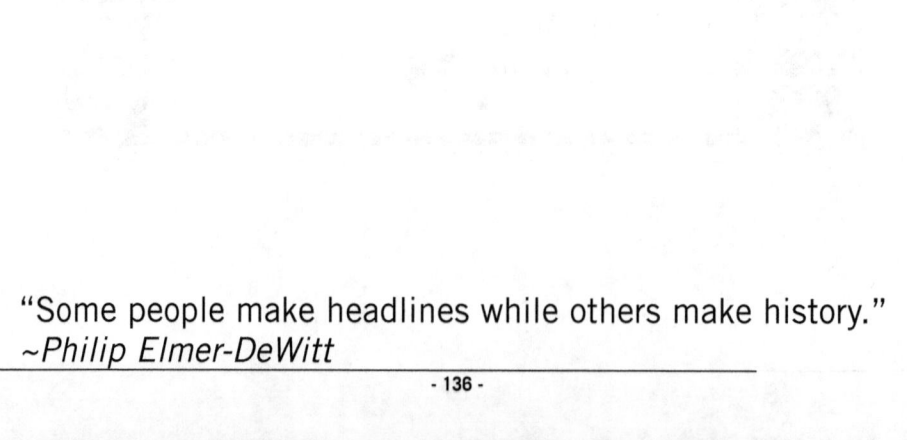

"Some people make headlines while others make history."
~*Philip Elmer-DeWitt*

REFERENCE

MANUALS
& INSTRUCTIONS

I<small>F YOU'VE LOST YOUR MANUAL, HELP IS ONLY A TEXT AWAY</small>

APPSPACE
iPhone app recommendations.
http://appspace.com

MANUALS ONLINE
Can't find that manual to your cell phone? Use this online resource.
http://cellphone.manualsonline.com

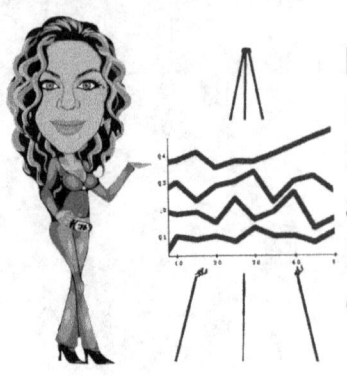

F.U.N.K.Y. STATISTIC:
Little Known Facts: First cell phone in the US is stated to be have been about 9 inches, by 5 inches by 1.75 inches weighed 2.5 pounds and was developed by Martin Cooper.

FOOD
FIND RECIPES AND THE BEST SEAFOOD ON THE GO

BLUE OCEAN INSTITUTE SEAFOOD GUIDE
Make seafood choices and learn about ocean conservation.
http://www.blueocean.org/seafood/seafood-guide
Type the words "TEXT" and your question about seafood to a
designated code. Receive an answer via SMS.

EPICURIOUS
Research recipes, create shopping lists and share with family
or friends.
Applicable Devices: iPhone
http://www.epicurious.com/services/mobile

WEBER GRILL
For you barbecue lovers, this is great for finding a grill, recipe,
meal planning, food preparation and even safety tips.
Applicable Devices: iPhone
http://www.weber.com/onthegrill
http://www.Weber.com

FOOD, ALLERGY & COLOUR ALERTS
Get the latest product recalls, withdrawals, and allergy alerts.
TEXT "START FOOD" to the number "62372".

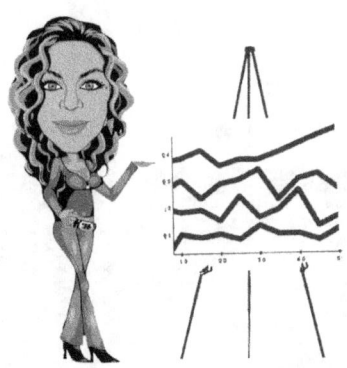

F.U.N.K.Y. STATISTIC:
Americans throw
out 27% of the
350,000,000
pounds of food
they buy each year.

LAW
RESEARCH THE LAW AND SO MUCH MORE

APP SHOPPER
Dictionary of legal and law terms - download app.
Applicable Device: iPhone
http://appshopper.com/reference/dictionary-of-legal-and-law-terms

BLACK'S LAW DICTIONARY
Definitive legal resource.
Applicable Device: iPhone
http://www.blackslawdictionary.com

LEGAL EDGE
Legal information right in your hands!
Applicable Device: iPhone
http://www.jdsupra.com/resources/syndication/iPhone.aspx

UT LAW
The University of Texas law library.
Applicable Devices: iPhone, iPOD Touch
WAP (Mobile WEBSITE) http://www.utexas.edu/law/m/
TWITTER @UTexasLaw

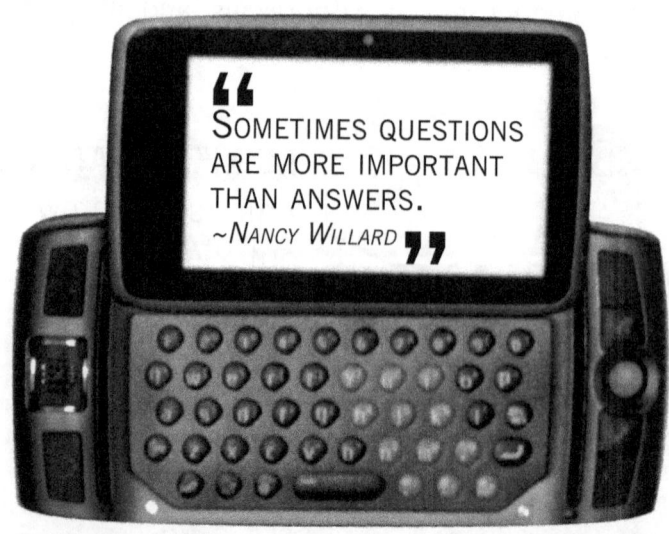

"
SOMETIMES QUESTIONS ARE MORE IMPORTANT THAN ANSWERS.
~*NANCY WILLARD* **"**

ANIMALS

FROM FEEDING TO FINDING YOUR PET TO WATCHING BIRDS - ALL ABOUT ANIMALS ON YOUR PHONE

AUDUBON
Bird Guide with GPS bird sightings, interactive information, field guides, bird sounds/calls, and you can share your photos and information with the world.
Applicable Device: iPhone
http://www.audubonguides.com

LOST MY DOGGIE
This is a service that helps individuals find lost or stolen pets.
TELEPHONE 1-877-818-0060

PETSMOBILITY
Pet tracking system. You can even set up geo-fences.
http://www.petsmobility.com/products/pawTrax.php

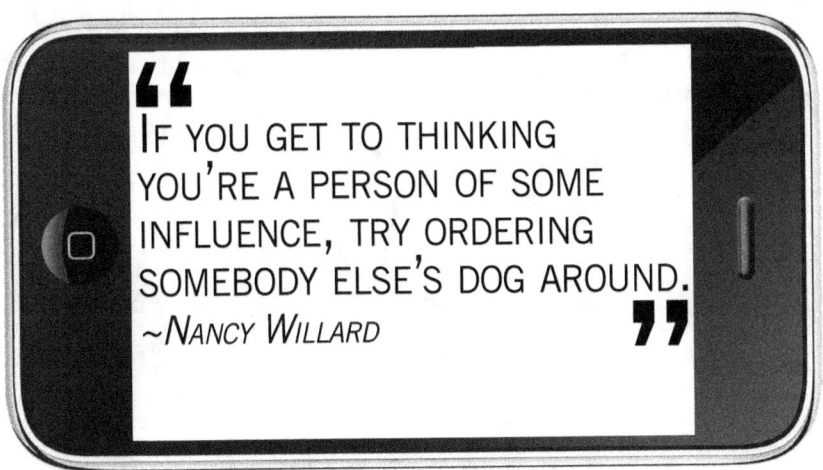

IF YOU GET TO THINKING YOU'RE A PERSON OF SOME INFLUENCE, TRY ORDERING SOMEBODY ELSE'S DOG AROUND.
~NANCY WILLARD

AT THE SPEED OF LIGHT

by Marrico Gordon

What has technology done for us?
Frustrations of a busy life
Now reduced.
An alerting reminder plays her favorite tune,
One appointment at ten, another at noon.
She's able to complete her project
As she flies back to Berkley.
With greater access to information,
She may finish without hurrying.
30,000 feet below, a foreman chirps out instructions,
"We're going to start laying cement north of here."

Saving money and time paying bills online,
Everything is made conveniently available.
Just when you think you don't have room to breathe,
Driving by snowy mountains headed east,
A navigation device puts your nerves at ease.

Thanks to his scheduler
He was able to remember his in-laws' anniversary -
50 years of commitment, one year of planning ahead,
Electronic memories so his mind won't forget.
Running to and fro,
Carrying a briefcase and his notebook on a rainy day,
He rushes into Starbucks and sets up an office-like display.
This is business, and, as usual,
A businessman rarely has time
For family, for friends, for himself.

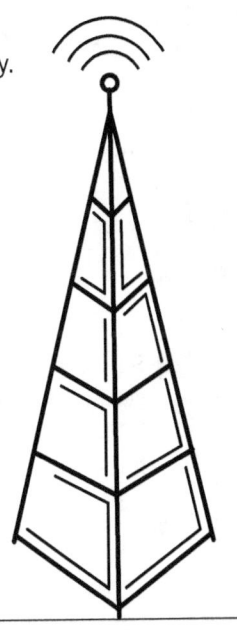

Finally granted a moment's peace.
Resting in bed with his laptop,
sipping hot coffee.
A trip to Suncadia sounds nice,
He makes the reservations through his mobile device.

What has technology done for me, you ask?
It has allowed me to travel
Beyond the seams of captivity,
Of inclusiveness.
Our soulless adventure radically inclined.

ADDITIONAL TIPS

ASSISTIVE TECHNOLOGY
For those with vision disabilities, consider getting a Jitterbug cell phone. They have large buttons which makes dialing much easier. You can even call the 24hour customer service line to have a number dialed on your behalf AND your phone book updated.
http://www.jitterbug.com/phones

PHONE SERVICE
Make certain a cell phone you purchase outside of buying from a service provider is NOT stolen. You can check to see if a cell phone's MEID (See Glossary) has been reported lost or stolen. You can do this by calling the service provider such as AT&T, T-Mobile, Verizon, Sprint. You can then ask them to verify the status of the cell phone's MEID of the cell phone. If you are able to activate the service for the cell phone, it is most likely you have a green light to purchase the cell phone, especially if refurbished. But if the cell phone's MEID has been reported lost or stolen I would advise NOT to purchase the cell phone.

WELLNESS
Were you aware that cell phones emit electro magnet waves that are known to be harmful to humans? There is a handy fashionable tool you can use called a "Ladybug" provided by Zerofon. You apply this decorative item directly on your cell phone using the adhesive that is provided. This nifty item helps to neutralize the electro magnetic waves that are emitted from your cell phone.http://budurl.com/LadyBug

If you would like to know who those annoying pesky calls come from that do not display a caller ID, try this, press the keys *#30# on your cell phone to have the caller ID displayed.

Say you are on the phone chatting with someone and then another call comes in...you then look for the switch call button to answer the other line. Next time try this little trick...press and hold the # key to toggle back and forth between the two lines.

Here is a preventative tip for you. If your cell phone is ever lost or stolen, you can lock it so that you don't have additional charges for usage you did not incur. Press the keys *#06# on your cell phone. Your cell phone will display a 15 digit serial number which is your IMEI (See Glossary). Keep these digits handy. If your cell phone is ever lost or stolen, you can notify your service provider and provide this IMEI code. Your service provider can then disable your cell phone EVEN if the thief changes the SIM Card (See Glossary). This will help to prevent additional charges that you did not make in the event your cell phone is lost or stolen.

Here is a trick to use emergency battery power. On your dial pad of your cell phone, type *#3370#. This will kick in the reserved emergency power of your cell phone.
Note: This may not work on all cell phones. There is a trick to check if it will by typing the code *#4720#.

Here are some things to consider when shopping for a new cell phone:
* Does the cell phone provide all of the features I need?
* What warranty or guarantees do I obtain with the cell phone purchase?
* What is the return policy?
* Is the price reasonable and competitive with the market value?
* Am I willing to stick to a particular carrier for two years?
* What is the monthly budget for the cell phone I can commit to?
* Think of the top three MUST HAVES I need on a day-to-day basis.

Tired of sitting through a long drawn out message when waiting to leave a voicemail to a contact? Here is a way that may help you skip directly to leaving a voicemail. Press the appropriate keystroke on your dial pad of your cell phone applicable to your service provider:
  * Cingular OR Verizon - Press *
  * Sprint - Press 1
  * T-mobile - Press #

"If you were going to die soon and had only one phone call you could make, who would you call and what would you say? And why are you waiting?"
~Stephen Levine

GLOSSARY

USEFUL TERMS

1G - First generation of mobile wireless technology.

2G - Second-generation of mobile wireless telephone technology (cellular) with the introduction of data services including text messages (SMS).

3G - Third Generation mobile wireless technology that delivers increased speed, capacity and more over its predecessors. new wireless standard promising increased capacity and high speed data applications up to two megabits.

4G - Fourth generation of mobile wireless telephone technology. 4G is the successor to the 3G technology.

711 - The nationwide number for relay service TRS.

802.11 - IEEE standard for wireless networks.

911 - US emergency number.

999 - UK emergency number.

AIRTIME - Actual time talking on a cell phone. Note: Cell phone carriers offer reduced rates for off-peak usage.

APPS - Applications

ANDROID - Google's smartphone based on the Linux platform introduced in 2008. Also known as G1.

BINARY RUNTIME ENVIRONMENT FOR WIRELESS (BREW) - A open source development platform for creating mobile cell phone applications most likely for CDMA mobile devices.

BLUETOOTH - Wireless network that utilizes personal area network (PAN) standard to enable data connections between electronic devices within a 30-foot range.

BROADBAND - Describes the bandwidth (capacity) for digital technology to carry voice, video or data channels simultaneously.

®CARRIER - A service provider or operator for a communications company such that provides a cellular service plan for a wireless phone.

CODE DIVISION MULTIPLE ACCESS (CDMA) - A cellular technology that supports SMS with a message length of 120 characters. CDMA is widely used in North America.

DIGITAL - A technique that encodes data and information using a binary code of 0s and 1s of electrical pulses.

ELECTROMAGNETIC FIELD - An area containing electromagnetic energy.

ELECTROMAGNETIC RADIATION - Waves of electromagnetic energy.

ELECTRONIC SERIAL NUMBER (ESN) - A unique identification number embedded in a cell phone.

ENHANCED 911 (E911) - When a 911 call is placed along with information such as automatic number identification of the cell phone's cellular number and location information to the 911 operator.

ENHANCED DATA FOR GSM EVOLUTION (EDGE) - A GSM standard that uses existing GSM infrastructure.

FEDERAL COMMUNICATIONS COMMISSION (FCC) - The Federal government agency that regulates telecommunications in the US.

GENERAL PACKET RADIO SERVICE (GPRS) - Boosts wireless data transmission over GSM networks.

GLOBAL POSITIONING SYSTEM (GPS) - A system of 24 satellites in orbit above the earth's atmosphere used to identify points of interest on the earth. locations, launched by the US By triangulation of signals from three of the satellites, a receiving unit can pinpoint its current location anywhere on earth to within a few meters.

HANDHELD DEVICE MARKUP LANGUAGE (HDML) - A version of the HTML language used for cell phones and Smartphones to view and obtain information on a website.

HANDS-FREE - A safety technique that allows a cell phone user to talk on a cell phone without permitting them to hold the phone to their ear. This greatly assists with minimizing car accidents of drivers talking on a cell phone while driving.

HERTZ (HZ) - A speed of unit measured by one cycle per second. The higher the number the faster the cycles per second.

INFRARED DATA ASSOCIATION (IRDA) - A port on a laptop, cell phone, PDA, or Smartphone in which data can be exchanged without a wired connection.

INTEGRATED CIRCUIT CARD ID (ICCID) - 19 or 20-digit serial number of a SIM card used in a cell phone.

INTERNATIONAL MOBILE EQUIPMENT IDENTIFIER (IMEI) - A serial number that has a unique 15-digit number located on the back of a cell phone.

INTERNATIONAL TELECOMMUNICATION UNION (ITU) - An agency of the United Nations that governs and develops telecommunications services worldwide.

INTEROPERABILITY - The ability of a device to operate with other devices or networks using different protocols or technologies.

LAND LINE - Traditional wired telephone.

LITHIUM-ION BATTERY (LI-ION) - A newer type of battery used in cell phones or devices that weighs less than its predecessors. Also known to last longer.

LG - A Korean company that manufactures cell phones and other devices. LG stands for "Life's Good".

MEGAHERTZ (MHZ) - The cycles per second of a unit of frequency equivalent to a million hertz per second. The higher the number the faster the cycles per second.

MOBILE SATELLITE SERVICE - Communications transmission service provided by satellites.
Note: A single satellite can provide coverage to a whole continent.

MULTIMEDIA MESSAGING SERVICE (MMS) - A cell phone message that contains an of an attachment containing multimedia such as a picture, video or sound.

NATIONAL EMERGENCY NUMBERING ASSOCIATION (NENA) - A mission to implement a universal emergency telephone number system.

NORTH AMERICAN CELLULAR NETWORK (NACN) - An organization that facilitates cell phone calls in the US.

NUMBER PORTABILITY - A service provided where consumers are able to keep their existing cell phone numbers when they change cell phone carriers.

OPERATING SYSTEM (OS) - The user interface and system that receives the input from a user interface and displays information to a screen. Examples are ¬† Symbian, Windows Mobile, and Palm and many more.

OTA - Over The Air wireless networking technology.

OVER-THE-AIR SERVICE PROVISIONING (OTASP) - The ability of a service provider or carrier to add or revise services to a cell phone over the air.

PCS Phone - PCS stands for portable communication system supported mostly by GSM.

PERSONAL DIGITAL ASSISTANT (PDA) - A mobile device that provides functions such as E-mail, calendar, calculator and more.

PUBLIC SWITCHED TELEPHONE NETWORK (PSTN)
Traditional Telephone system.

PULL SMS - The ability to pull data such as ringtones or other Apps from a cell phone.

PUSH SMS - The ability to request data from a cell phone.

RADIATION - The emission of energy.

RADIO FREQUENCY (RF) - Electro-magnetic energy between audio and light.

ROAMING - The ability to use a cell phone outside a coverage area.

SERVICE CHARGE - The cost of service paid for cell phone service paid either a one time fee or on a monthly basis.

SERVICE PLAN - The service you pay for the usage of a cell phone. This may included SMS service, a data plan for accessing the internet etc.

SIM CARD - A chip made of plastic used in cell phone that contains the cell phone's IMEI, ICCID, service plan, cell phone number.

SKYPE - A service that uses software to allow a user to place a telephone call over the Internet, thus lowering the cost of a long-distance phone call. Skype now incorporates SMS, video conferencing, chat and more.

SMART PHONE - A cell phone with advanced features that can perform functions similar to a desktop computer.

SPECIFIC ABSORPTION RATE (SAR) - A rate at which radio frequency is measured and absorbed by the human body.

T9 - Which means "Text on 9 keys". T9 is a way to send SMS messages to a cell phone using a predictive text technology.

TALK TIME - Same as "Talk Time". The time you have to talk on a cell phone without having to recharge your battery.

TELECOMMUNICATIONS - The transmission of communication that includes text, music, or photos.

TERMINATION CHARGES - Charges a service provider will charge an individual for terminating a cell phone plan earlier than the contract was slated for.

TRI BAND - A cellular network that operates in three frequency bands (800 MHz, 900 MHz and 1800MHz).

TRS - Provides a service to people with speech or hearing disabilities.

TWEET - A message of no more than 140 characters sent or received to a micro-blogging service called Twitter (See Twitter).

TWITTER - A social media and micro-blogging service.

UPLINK - Communications from a device or computer to a satellite.

VOICE ACTIVATION - A method of communication that uses voice to activate a function on a cell phone such as dialing a contact.

VOICE RECOGNITION - The ability of a cell phone to recognize a voice command used to carry out a function.

WIRELESS APPLICATION PROTOCOL (WAP) - A mobile version of HTML used to access the internet on a cell phone using the Wireless Application Protocol.

WI-FI - ALSO KNOWN AS "802.11". A wireless standard allows a user to connect to the internet or a device without any wired connections.

WIRELESS INSTANT MESSAGING (WIM) - A service that allows a computer or laptop user to send a message to a cell phone.

"Hide not your talents, they for use were made. What's a sun-dial in the shade?"
~ *Benjamin Franklin*

EMOTICONS

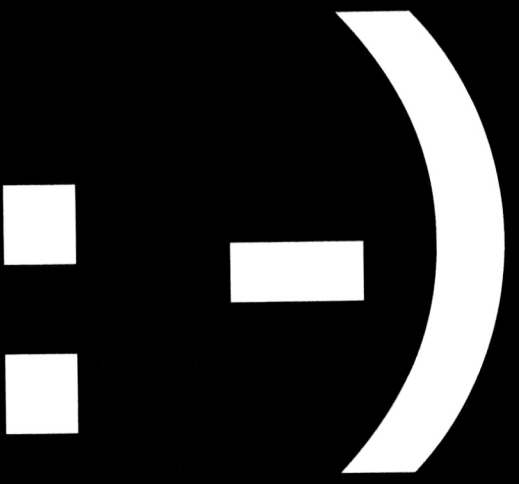

SYMBOL	MEANING
2	TOO
4	FOR
9	PARENT WATCHING
99	PARENT IS NO LONGER WATCHING
$_$	GOT MONEY
((HUG))	HUG
G	GRIN
O	SHOCKED
-.-	ANNOYANCE
:$	EMBARRASSED
:(	SAD
:'(	CRYING
:-(	SAD
:)	HAPPY
:-)	SMILE
:-*	KISS
:/	SARCASM
:?	CONFUSED
:@	ANGRY
:]	HAPPY
:I	DISGUST
:0	SURPRISED
:3	GOOFY
:9	YUMMY
:C	VERY SAD
:D	HAPPY
:F	DROOLING
:L	LAUGHING
:O	SURPRISED

:P	STICKING TONGUE OUT
:S	CONFUSED
:X	MY LIPS ARE ZIPPED
;-)	WINK
;D	WINK
;O	JOKING
;P	WINKING & STICKING TONGUE OUT
?	WHAT?
@	AT
@_@	DAZED
^.^	HAPPY
^^	HAPPY
^_^	HAPPY
^5	HIGH FIVE
-_-	ANNOYED, TIRED
<_<	SARCASTIC LOOK
<>	NO COMMENT
<><	FISH
<G>	GRIN
=.=	TIRED
=]	HAPPY
=D	:D
=S	CONFUSED
=X	NO COMMENT
0.0	SURPRISE
O:)	INNOCENT
O_O	CONFUSED
_	IN LOVE, DAZED
%)	DRUNK, GIDDY

"I don't like people who take drugs...Customs men for example."
~*Mick Miller*

ACRONYMS

LOL

SYMBOL	MEANING
2BH	TO BE HONEST
2DAY	TODAY
2EZ	TOO EASY
2G2BT	TOO GOOD TO BE TRUE
2K10	2010
2L8	TOO LATE
2M	TOMORROW
2MI	TOO MUCH INFORMATION
2MRO	TOMORROW
3Q	THANK YOU
4COL	FOR CRYING OUT LOUD
4RL?	FOR REAL?
4SHO	FOR SURE
4U	FOR YOU
4YEO	FOR YOUR EYES ONLY
5N	FINE
ABT	ABOUT
AFAIK	AS FAR AS I KNOW
AIIC	AS IF I CARE
ANY1	ANYONE
ANYWHO	ANYHOW
AYKM	ARE YOU KIDDING ME?
B/C	BECAUSE
B4	BEFORE
B4N	BYE FOR NOW
BB	BE BACK
BBL	BE BACK LATER
BBS	BE BACK SOON
BDAY	BIRTHDAY
BF	BOYFRIEND

BF4L	BEST FRIENDS FOR LIFE
BIAB	BACK IN A BIT
BRB	BE RIGHT BACK
BRT	BE RIGHT THERE
BTD	BORED TO DEATH
BTDT	BEEN THERE DONE THAT
BTS	BE THERE SOON
BTW	BY THE WAY
BYO	BRING YOUR OWN
C	SEE
CIO	CHECK IT OUT
CM	CALL ME
CML	CALL ME LATER
COO	COOL
CSL	CAN'T STOP LAUGHING
CTN	CAN'T TALK NOW
CU	SEE YOU
CUL	SEE YOU LATER
CWYL	CHAT WITH YOU LATER
CYA	SEE YA
CYE	CHECK YOUR EMAIL
CYL	CATCH YOU LATER
CYL8R	SEE YA LATER
CYM	CHECK YOUR MAIL
D8	DATE
DAT	THAT
DC	DON'T CARE
DK	DON'T KNOW
DMAL	DROP ME A LINE
DNO	DON'T KNOW

HELPFUL LINKS

Here are some additional resources to obtain mobile apps for your mobile cell phone.

Covers most mobile devices (iPhone, Blackberry, Palm, Android, And Windows Mobile):
* http://www.handango.com
* http://www.youpark.com
* http://www.google.com/mobile
* http://download.cnet.com/mobile
* http://www.mymobilewebapps.com
* http://www.getjar.com
* http://mplayit.com
* http://mobileiron.com
* http://www.shozu.com/portal/index.do

BLACKBERRY
* http://www.engadget.com/tag/AppWorld
* http://BlackBerryMobile.com
* http://appworld.blackberry.com/webstore
* http://www.blackberrycool.com
* http://www.squidoo.com/blackberry-apps
* http://www.blackberryfreeware.com
* http://na.blackberry.com/eng
* http://www.berryreview.com/blackberry-software-directory
* http://crackberry.com/2-99-minimum-paid-app-price-blackberry-app-world
* http://mobile.softwareload.co.uk for those in the UK (from a PC) WAP http://m.softwareload.com (from a mobile device).
* http://www.bbgeeks.com/blackberry-applications/the-best-in-free-blackberry-software-8897
* http://www.blackberryfreeware.com
* Free BlackBerry Themes - www.themes4bb.com

* http://gadgetsteria.com/2009/03/31/3-blackberry-app-store-com
  parisons-app-world-crackberry-mobihand

APPLE IPAD
* http://www.apple.com/ipad/apps-for-ipad

APPLE IPHONE
* http://www.ondeego.com. Great for those of you desiring to
manage mobile devices in the workplace! To join the beta
go to http://www.theappcentral.com.
* http://www.148apps.com
* http://www.iphonestop25.com
* http://www.iphonegizmo.com
* http://www.appsmile.com
* http://appadvice.com/appbase
* http://www.apple.com/iphone/apps-for-iphone

APPLE IPOD
* http://www.applerepo.com
* http://www.topitouchapps.com

ANDROID
* http://101bestandroidapps.com
* http://www.androidtapp.com
* http://www.android.com/market
* http://www.androlib.com/android.category.applications-j.aspx
* http://androidandme.com/category/applications/

PALM
* http://www.palmgear.com/en/usd/index.html
* http://www.freeware-palm.com

* http://www.palminfocenter.com
* http://www.palmgear.com
* http://Freeware.Pronto.com

NOKIA
* http://apps.store.ovi.com
  WAP - store.ovi.com

WINDOWS MOBILE
* http://pocketgear.com
* http:/handango.com
* http://www.1800pocketpc.com
* http://marketplace.windowsphone.com/Default.aspx
* http://www.wm6software.net
* http://www.smartphonefreeware.org
* http://www.windowsmobilesoft.net

CITED RESOURCES

ADDITIONAL TIPS

Tips for safer cell phone use. (n.d.). Retrieved from http://www.virginiahopkinstestkits.com/safercellphone.html

ASSISTIVE TECHNOLOGY

Miller, JM. (2009). savvy senior::simple cell phone options for seniors. The Charleston Gazette, Retrieved from http://ezproxy.spl.org:2048/login?url=http://proquest.umi.com/pqdweb?did=164298 0411&Fmt=3&clientId=11206&RQT=309&VName=PQD doi: 1642980411

Anonymous, Initials. (2009). Vlingo 4.0 Plus Featuring 'Vlingo Everywhere' Extends Mobile Voice Features to Virtually Every Application on BlackBerry Smartphones; Users have now used Vlingo over 100,000,000 times. Pr newswire. Retrieved (2009, December 23) from http://ezproxy.spl.org:2048/login?url=http://proquest.umi.com.ezproxy.spl.org:2048/pqdweb?did=1918417491&sid=1&Fmt=3&clientId=11206&RQT=309&VName=PQD

Miller, JM. (2009). savvy senior::simple cell phone options for seniors. The Charleston Gazette, Retrieved from http://ezproxy.spl.org:2048/login?url=http://proquest.umi.com/pqdweb?did=164298 0411&Fmt=3&clientId=11206&RQT=309&VName=PQD doi: 1642980411818-569-3057

DATA RECOVERY

Cellebrite usa releases the must-have application for mobile retailers; transfers phone book contacts and multimedia content between cell phones at the touch-of-a-button. (2006). Business Wire, Retrieved from http://ezproxy.spl.org:2048/login?url=http://proquest.umi.com/pqdweb?did=1018962921&Fmt=3&clientId=11206&RQT=309&VName=PQD doi: 1018962921

SHAMAH, DS. (2009). Organization, cellphone style. The Jerusalem

Post, Retrieved from http://ezproxy.spl.org:2048/login?url=http://proquest.umi.com/pqdweb?did=1897050451&Fmt=3&clientId=11206&RQT=309&VName=PQD doi: 1897050451

Anonymous, Initials. (2009, June 23). M2 presswire. Firefox SMS Systems: Firefox(R) SMS Systems Introduces Saviour for Cell Phone

Contacts; SaveCell(R) can remotely remove phonebook contents from stolen phones, Retrieved from http://ezproxy.spl.org:2048/login?url=http://proquest.umi.com/pqdweb?did=1757155571&Fmt=3&clientId=11206&RQT=309&VName=PQD doi: 1757155571

SOCIAL MEDIA

Lowensohn, JL. (2009, December 21). Video site dailymotion gets a pair of iphone apps. Retrieved from http://reviews.cnet.com/iphone-atlas/?keyword=apps

Fisher, AF. (2009, August 24). Flickr. Retrieved from http://www.time.com/time/specials/packages/article/0,28804,1918031_1918016_1918028,00.html

Morford, MM. (2007). Wanna hook up? let your thumbs do the dialing. San Francisco Chronicle, (19328672), Retrieved from http://ezproxy.spl.org:2048/login?url=http://proquest.umi.com/pqdweb?did=1253898401&Fmt=3&clientId=11206&RQT=309&VName=PQD doi: 1253898401

Easems midlet 1.0. (2006, December 31). Retrieved from http://www.filetransit.com/view.php?id=33597

Anonymous, Initials. (2008, August 13). Fishtext: fishtext: more than 100,000 users in three weeks; new application for cheap sms text messaging offers big savings over pay as you go tariffs. M2 Presswire.

Mann, MM. (2009). 13 mobile apps that work - presentation transcript. Retrieved from http://www.slideshare.net/marinamann/13-mobile-apps-that-work

Anonymous, Initials. (2009, September 7). e-world: who says there are no free lunches?. Businessline, Retrieved from http://ezproxy. spl.org:2048/login?url=http://proquest.umi.com/pqdweb?did=185 4636751&Fmt=3&clientId=11206&RQT=309&VName=PQD doi: 1854636751

Anonymous, Initials. (2009). 'kyodo news' available for iphone/ipod touch through 'kijizo' news distribution system developed by east. PR Newswire, Retrieved from http://ezproxy.spl.org:2048/login?url=http:// proquest.umi.com/pqdweb?did=1875499351&Fmt=3&clientId= 11206&RQT=309&VName=PQD doi: 1875499351 Anonymous, Initials. (2008). Loopt Announces Free Application for Apple App Store:Connecting Users to the World Around Them Like Never Before. Pr newswire. Retrieved (2009, December 23) from http://ezproxy.spl. org:2048/login?url=http://proquest.umi.com/pqdweb?did=15081790 01&Fmt=3&clientId=11206&RQT=309&VName=PQD

Fisher, AF. (2009, August 24). 50 best websites 2009. Retrieved from http://www.time.com/time/specials/packages/article/0,28804,1918031_1918016_1917971,00.html

Mann, MM. (2009). 13 mobile apps that work - presentation transcript. Retrieved from http://www.slideshare.net/marinamann/13-mobile-apps-that-work http://www.drivinglaws.org/washington.php

LITERATURE
Bradbury, DB. (2009, June 30). The Ideal task manager. National Post, 14868008.

MUSIC
Appendix:
AnneH, AH. (n.d.). Information on copyright laws on cds. Retrieved from http://www.ehow.com/facts_6031204_information-copyright-laws-cds.html

Sony music entertainment websites to feature audio players and song lyrics . (2009, May 05). PR Newswire, Retrieved from http:// ezproxy.spl.org:2048/login?url=http://proquest.umi.com/pqdweb?d id=1699032181&Fmt=3&clientId=11206&RQT=309&VName=P

QD doi: 1699032181.

Sony music entertainment websites to feature audio players and song lyrics . (2009, May 05). PR Newswire, Retrieved from http://ezproxy.spl.org:2048/login?url=http://proquest.umi.com/pqdweb?did=1699032181&Fmt=3&clientId=11206&RQT=309&VName=PQD doi: 1699032181.

Anonymous, Initials. (2008, October 16). Gracenote and metrolyrics extend authorized lyrics service to mobile devices and iphone applications. Canada NewsWire, Retrieved from http://ezproxy.spl.org:2048/login?url=http://proquest.umi.com/pqdweb?did=157 810351&Fmt=3&clientId=11206&RQT=309&VName=PQD doi: 1575810351.

Sony music entertainment websites to feature audio players and song lyrics . (2009, May 05). PR Newswire, Retrieved from http://ezproxy.spl.org:2048/login?url=http://proquest.umi.com/pqdweb?did=1699032181&Fmt=3&clientId=11206&RQT=309&VName=PQD doi: 1699032181.

Music to your ears. (2006). http://www.mobilemarketingmagazine.co.uk , Retrieved from http://www.mobilemarketingmagazine.co.uk/2006/05/music_to_your_e.html

Sony music entertainment websites to feature audio players and song lyrics . (2009, May 05). PR Newswire, Retrieved from http://ezproxy.spl.org:2048/login?url=http://proquest.umi.com/pqdweb?did=1699032181&Fmt=3&clientId=11206&RQT=309&VName=PQD doi: 1699032181.

Sony music entertainment websites to feature audio players and song lyrics . (2009, May 05). PR Newswire, Retrieved from http://ezproxy.spl.org:2048/login?url=http://proquest.umi.com/pqdweb?did=1699032181&Fmt=3&clientId=11206&RQT=309&VName=PQD doi: 1699032181.

Txtdrop.com news & press releases . (2009, October 18). Retrieved from http://www.txtdrop.com/news.php

Anonymous, Initials. (2009, July 21). Namm launches free 'wanna play music' mobile application on itunes, game app tests players' 'musicality' and features music lessons locator:free music game app from namm available now in itunes store . PR Newswire, 1797969461.

Kemp, LK. (2009, August 23). Introducing rhapsody on the iphone. Retrieved from http://realnetworksblog.com/?p=889

VIDEO
Martin, DM. (2009, December 17). Eyetv 3.3 conquers 3g iphone video streaming. Retrieved from http://reviews.cnet.com/iphone-atlas/?keyword=apps

SPORTS
Marazita, FM. (2009). Need a gift for your golfer. Linkedin. Retrieved (2009, December 22) from http://www.linkedin.com/group Answers?viewQuestionAndAnswers=&discussionID=11097146&gid=96032&trk=EML_anet_qa_ttle-cThOonOJumNFomgJt7dBpSBA

Anonymous, Initials. (2009, September 25). Sprint reaches nfl fans through creative forms of marketing outreach efforts. Business Wire, Retrieved from http://ezproxy.spl.org:2048/login?url=http://proquest.umi.com/pqdweb?did=1866452021&Fmt=3&clientId=11206&RQT=309&VName=PQD doi: 1866452021

GAMES
Rei, JR. (2009). Magnetic Sports Soccer is now available on iPhone and iPod Touch.. Linkedin. Retrieved (2009, December 22) from http://www.linkedin.com/groupAnswers?viewQuestionAndAnswers=&discussionID=11072830&gid=96032&trk=EML_anet_qa_ttle-cThOonOJumNFomgJt7dBpSBA

Belic, DB. (2009). GameHouse's COLLAPSE! casual game now available across multiple platforms – BlackBerry, iPhone and Android included Read more: http://www.intomobile.com/2009/12/12/gamehouses-collapse-casual-game-now-available-across-multiple-platforms-blackberry-iphone-and-android-included.html#ixzzOaUAxhqgS.

Intomobile. Retrieved (2009, December 22) from http://www.intomo-bile.com/2009/12/12/gamehouses-collapse-casual-game-now-avail-able-across-multiple-platforms-blackberry-iphone-and-android-includ-ed.html

Anonymous, Initials. (2009). Gosub 60 makes history with the first ebook fused mobile game; new sherlock holmes game combines gameplay with classic stories. PR Newswire, Retrieved from http://ezproxy.spl.org:2048/login?url=http://proquest.umi.com.ezproxy.spl.org:2048/pqdweb?did=1917643461&sid=1&Fmt=3&clientId=11206&RQT=309&VName=PQD doi: 1917643461

Mann, MM. (2009). 13 mobile apps that work - presentation tran-script. Retrieved from http://www.slideshare.net/marinamann/13-mobile-apps-that-work

BARCODE

Sign up today for best buy mobile alerts: 33221. (n.d.). Retrieved from http://stores.bestbuy.com/1410/2009/06/17/sign-up-today-for-best-buy-mobile-alerts-332211.

I-nigma barcode reader. (2009, August 16). Retrieved from http://www.freewarepocketpc.net/ppc-download-i-nigma-barcode-reader-v1-4.html

Simon, . (2009, December 28). we're back! semacode + iphone = free qr code reader :-). Retrieved from http://semacode.com

PHONE SERVICE

Anonymous, Initials. (2009). idt introduces pennytalk(r) mobile app for iphone(r) users to save big on international calls. Business Wire, Retrieved from http://ezproxy.spl.org:2048/login?url=http://proquest.umi.com/pqdweb?did=1906962751&Fmt=3&clientId=11206&RQT=309&VName=PQD doi: 1906962751

CATALOGS

Hepburn, AH. (2009). IKEA iPhone App: The 2010 Catalogue. Dig-italbuzz blog. Retrieved (2009, December 22) from http://www.dig-italbuzzblog.com/ikea-iphone-app-the-2010-catalogue

COUPONS & SALES

Sign up today for best buy mobile alerts: 33221. (n.d.). Retrieved from http://stores.bestbuy.com/1410/2009/06/17/sign-up-today-for-best-buy-mobile-alerts-332211.
SMS 1 TEXT keyword "GAMER" to shortcode "332211"
SMS 2 For help, TEXT "HELP" to 332211 anytime.
SMS 3 To cancel, TEXT "STOP" to 332211 anytime.

Anonymous, Initials. (2008, September 8). Finding the cheapest local gas in the palm of your hand:premium sms company offers gas alerts via sms . PR Newswire, Retrieved from http://ezproxy.spl.org:2048/login?url=http://proquest.umi.com/pqdweb?did=1550097491&Fmt=3&clientId=11206&RQT=309&VName=PQD doi: 1550097491 SMS 1 TEXT "gasbuddy" + zip code or city, state to "368266"

Anonymous, Initials. (2009, September 23). Goldspot media launches miapp, industry's first 'design once, deploy in any app store' mobile app creation platform:single web interface to create, manage and maintain apps, advertising and promotions across all mobile devices. PR Newswire, Retrieved from http://ezproxy.spl.org:2048/login?url=http://proquest.umi.com/pqdweb?did=1864386641&Fmt=3&clientId=11206&RQT=309&VName=PQD   doi: 1864386641

Anonymous, Initials. (2009, March 19). Shortcuts.com introduces new shortcuts.com mobile. Business Wire, Retrieved from http://ezproxy.spl.org:2048/login?url=http://proquest.umi.com/pqdweb?did=1663942531&Fmt=3&clientId=11206&RQT=309&VName=PQD doi: 1663942531.

Dilworth, DD. (2009). valpak opens up mobile local couponing with iphone application. DM News, (01943588), Retrieved from http://ezproxy.spl.org:2048/login?url=http://proquest.umi.com/pqdweb?did=1885044091&Fmt=3&clientId=11206&RQT=309&VName=PQD doi: 1885044091.

AUCTIONS

Anonymous, Initials. (2009). Wapple: CeX launches new second hand 'phone trading service via Wapple powered mobile web service

- turn your old phone into cash now!. M2 presswire. Retrieved (2010, January 12) from http://ezproxy.spl.org:2048/login?url=http://pro-quest.umi.com/pqdweb?did=1917779271&Fmt=3&clientId=112 06&RQT=309&VName=PQD

Davies, PD. (2008, April 11). Christie's redesigns its website. Re-trieved from http://www.news-antique.com/?id=784151

FOOD

Coupons.com mobile application now available in apple app store. (2009, December 14). Retrieved from http://www.couponsinc.com/corp/source/oc_press_release.asp?art=20091211&yr=2009&t=R

Tsirulnik, GT. (2009, February 6). Fast-food chain subway launches mobile ordering system. Retrieved from http://www.mobilemarketer.com/cms/news/commerce/2593.html

REALESTATE

Anonymous, Initials. (2009, August 18). mobile app connects with new home buyers . Canada NewsWire, Retrieved from http://ezproxy.spl.org:2048/login?url=http://proquest.umi.com/pqdweb?did=183 4745531&Fmt=3&clientId=11206&RQT=309&VName=PQD doi: 1834745531 SMS 1 TEXT "VIEW" To "51945" for a demo

AUTOMOBILES

Anonymous, Initials. (2009, November 28). Txt2look inc.; txt2look expands into california & florida. Real Estate Business Journal, Re-trieved from http://ezproxy.spl.org:2048/login?url=http://proquest.umi.com/pqdweb?did=1904115561&Fmt=3&clientId=11206&RQ T=309&VName=PQD doi: 1904115561.
Anonymous, Initials. (2009). Kelley Blue Book's Kbb.com to Launch All-New Online Vehicle Classifieds: The Trusted Marketplace(SM). Pr newswire. Retrieved (2009, December 10) from http://ezproxy.spl.org:2048/login?url=http://proquest.umi.com/pqdweb?did=173 2994451&Fmt=3&clientId=11206&RQT=309&VName=PQD
Anonymous, Initials. (2008, April 4). Find your ideal mini. Daily Record, Retrieved from http://ezproxy.spl.org:2048/login?url=http://proquest.umi.com/pqdweb?did=1456484531&Fmt=3&clientId=1

1206&RQT=309&VName=PQD doi: 1456484531

Cupr, TC. (2009). Confused about cars? iPhone can help!. Linkedin. Retrieved (2009, December 22) from http://icardata.co.uk/car-
data-products/new-car-guide-2010

SPECIALTY

fossil unveils watch that integrates with cell phone; caller id in stylish watch helps you stay in touch and in 'style'. (2006). PR Newswire, Retrieved from http://ezproxy.spl.org:2048/login?url=http://
proquest.umi.com/pqdweb?did=1136582661&Fmt=3&clientId=1
1206&RQT=309&VName=PQD doi: 1136582661

BENEVOLENCE

Winslow, LW. (2009). Mobile app may benefit charities. Tri-
bune Business News, Retrieved from http://ezproxy.spl.org:2048/
login?url=http://proquest.umi.com/pqdweb?did=1885543691&Fm
t=3&clientId=11206&RQT=309&VName=PQD doi: 1885543691

Anonymous, Initials. (2009). Bringing the holidays to military men
and women. U.S. Newswire, Retrieved from http://ezproxy.spl.
org:2048/login?url=http://proquest.umi.com/pqdweb?did=187814
6981&Fmt=3&clientId=11206&RQT=309&VName=PQD
doi: 1878146981

iRecycle. (n.d.). Retrieved from http://www.appsafari.com/lo-
cal/9261/irecycle DURANDO, JD. (2009). Cell phone texting makes
giving as easy as 1-2-3. Retrieved from http://ezproxy.spl.org:2048/
login?url=http://proquest.umi.com/pqdweb?did=1878440461&Fm
t=3&clientId=11206&RQT=309&VName=PQD doi: 1878440461.
SMS 1TEXT the word "Alive" to the number "90999."

Coulter, MC. (2009). iGIVE. City paper. Retrieved (2009, December 23) from http://ezproxy.spl.org:2048/login?url=http://proquest.
umi.com/pqdweb?did=1861887231&Fmt=3&clientId=11206&
RQT=309&VName=PQD Krishna, SK. (2010, January 14). Salvation army text to give to haiti:text haiti to 45678. Retrieved from
http://www.nowpublic.com/world/salvation-army-text-give-haiti-text-
haiti-45678-2556660.html

GOING GREEN

Richard, MGR. (2005, November 26). Treehugger homework: unplug your cellphone charger. Retrieved from http://www.treehugger.com/files/2005/11/treehugger_home_2.php

DeFreitas, SD. (2009). Eco-activism goes mobile with ecosnoop iphone app. EarthTechling, Retrieved from http://www.earthtechling.com/2009/12/eco-activism-goes-mobile-with-ecosnoop-iphone-app

Smartrecycle system. (n.d.). Retrieved from http://www.batteryrecycling.com/SmartRecycle+System

CRISIS

(2008). San Diego Union.

Wood, CW. (2009). American heart association pushes cpr tactic with social . DM News, Retrieved from http://ezproxy.spl.org:2048/login?url=http://proquest.umi.com/pqdweb?did=1904899041&Fmt=3&clientId=11206&RQT=309&VName=PQD doi: 1904899041

New iphone app - i am safe. (2009, September 29). Retrieved from http://topiphonenews.com/new-iphone-app-i-am-safe

Anonymous, Initials. (2009). Nationwide(r) mobile application now available on apple app store. Business Wire, Retrieved from http://ezproxy.spl.org:2048/login?url=http://proquest.umi.com/pqdweb?did=1684059031&Fmt=3&clientId=11206&RQT=309&VName=PQD doi: 1684059031

Morford, MM. (2007). Wanna hook up? let your thumbs do the dialing. San Francisco Chronicle, (19328672), Retrieved from http://ezproxy.spl.org:2048/login?url=http://proquest.umi.com/pqdweb?did=1253898401&Fmt=3&clientId=11206&RQT=309&VName=PQD doi: 1253898401 SMS 1 TEXT "sexinfo" to 61827

Sos1.tel: the one .tel domain that may save your life this holiday season . (2009, June 19). PR Newswire, Retrieved from http://ezproxy.spl.org:2048/login?url=http://proquest.umi.com/pqdweb?did=1751553291&Fmt=3&clientId=11206&RQT=309&VName=PQD doi: 1751553291.

MEDICAL

Fitness and health: cell-phone alert. (2009). Gannett News Service, Retrieved from http://ezproxy.spl.org:2048/login?url=http://proquest.umi.com/pqdweb?did=1882123951&Fmt=3&clientId=11206&RQT=309&VName=PQD doi: 1882123951

LaVallee, AL. (2009, October 5). App watch: a mobile swine-flu tracker. Wall Street Journal, Retrieved from http://blogs.wsj.com/digits/2009/10/05/app-watch-a-mobile-swine-flu-tracker

Mann, MM. (2009). 13 mobile apps that work - presentation transcript. Retrieved from http://www.slideshare.net/marinamann/13-mobile-apps-that-work

Anonymous, Initials. (2009). Deviceanywhere: healthcare industry turns to deviceanywhere to bring their products and services to the mobile platform; deviceanywhere gives developers peace-of-mind that mobile applications meet the rigorous standards of reliability and security required for mhealth. M2 Presswire, Retrieved from http://ezproxy.spl.org:2048/login?url=http://proquest.umi.com/pqdweb?did=1868147911&Fmt=3&clientId=11206&RQT=309&VName=PQD doi: 1868147911

Esp launches free mobile application for health care professionals. (n.d.). PR Newswire, Retrieved from http://ezproxy.spl.org:2048/login?url=http://proquest.umi.com/pqdweb?did=1883882571&Fmt=3&clientId=11206&RQT=309&VName=PQD doi: 1883882571.

Anonymous, Initials. (2009, November 11). Elsevier announces the launch of md consult mobile. PR Newswire, Retrieved from http://ezproxy.spl.org:2048/login?url=http://proquest.umi.com/pqdweb?did=1898431751&Fmt=3&clientId=11206&RQT=309&VName=PQD doi: 1898431751.

Anonymous, Initials. (2009). Medica first health plan to create mobile app for comparing health care costs . Business Wire, Retrieved from http://ezproxy.spl.org:2048/login?url=http://proquest.umi.com/pqdweb?did=1886703351&Fmt=3&clientId=11206&RQT=309&VName=PQD doi: 1886703351

Introducing new robitussin® to go!. (n.d.). Retrieved from http://www.4infoalerts.com/wap/robitussin.   SMS 1Receive alerts by signing up - TEXT "ROBITUSSIN" to "88398" SMS 2 You   may TEXT "ROBITUSSINSTOP" to unsubscribe.

Anonymous, Initials. (2009). Webmd launches free mobile application for physicians:medscape mobile provides the fastest, most comprehensive mobile medical information resource. PR Newswire, Retrieved   from   http://ezproxy.spl.org:2048/login?url=http://proquest.umi.com/pqdweb?did=1797131651&Fmt=3&clientId=11206&RQT=309&VName=PQD doi: 1797131651

NUTRITION
Anonymous, Initials. (2009). New mobile application from aetna helps college students stay fit . Business Wire, Retrieved from http://ezproxy.spl.org:2048/login?url=http://proquest.umi.com/pqdweb?did=1869402551&Fmt=3&clientId=11206&RQT=309&VName=PQD doi: 1869402551

SELF HELP
Tsirulnik, GT. (2009, December 23). Franklincovey launches career app. Retrieved from http://www.mobilemarketer.com/cms/news/content/4943.html

Mann, MM. (2009). 13 mobile apps that work - presentation transcript. Retrieved from http://www.slideshare.net/marinamann/13-mobile-apps-that-work

Anonymous, Initials. (2009, April 27). New campaign helps americans deal with stressful times. U.S. Newswire.
Broida, RB. (2009, December 9). New iphone apps aim to lower stress. Retrieved from http://reviews.cnet.com/8301-19512_7-10410810-233.html?tag=nl.e798

Beyer-Clausen, MBC. (2009, November 02). Virgin atlantic releases iphone app for people with a fear of flying.   Retrieved   from   http://www.linkedin.com/newsArticle?viewDiscussion=&articleID=82491182&gid=56468&trk=EML_anet_nws_title-cThOon0JumNFomgJt7dBpSBA.

PRODUCTIVITY

Chartier, DC. (2008). first look: alerts.com omninotification service needs polish (updated). Arstechnica, Retrieved from http://arstechnica.com/old/content/2008/06/first-look-alerts-com-omninotification-service-needs-polish.ars

Bradbury, DB. (2009, June 30). The Ideal task manager. National Post, 14868008.
Bradbury, DB. (2009, June 30). The Ideal task manager. National Post, 14868008.
Bradbury, DB. (2009, June 30). The Ideal task manager. National Post, 14868008.

Anonymous, Initials. (2009). Research and markets: microsoft's current and future mobile apps/services strategy in emerging markets. looking beyond oneapp. M2 Presswire, Retrieved from http://ezproxy.spl.org:2048/login?url=http://proquest.umi.com.ezproxy.spl.org:2048/pqdweb?did=1918251921&sid=1&Fmt=3&clientId=11206&RQT=309&VName=PQD doi: 1918251921

Anonymous, Initials. (2009, November 24). Iphone and ipod touch users log in and branch out with new ntrconnect remote access to business and home computers. U.S. Newswire. Washington: Nov 24, 2009.
Kiosk logix integrates printeron's remote printing services for business travelers.. (2005, July 25). Business Wire, (872176131), Retrieved from http://ezproxy.spl.org:2048/login?url=http://proquest.umi.com/pqdweb?did=872176131&Fmt=3&clientId=11206&RQT=309&VName=PQD doi: 872176131.

Bradbury, DB. (2009, June 30). The Ideal task manager. National Post, 14868008.

Abram, SA. (2007). You can take it with you: online apps help road warriors. Information Outlook, (10910808), Retrieved from http://ezproxy.spl.org:2048/login?url=http://proquest.umi.com/pqdweb?did=1389693021&Fmt=3&clientId=11206&RQT=309&VName=PQD doi: 1389693021

NETWORKING

Johnson, RJ. (2009). FREE iPhone Promo Codes for WorldCard Mobile Business Card Reader app!. Linkedin. Retrieved (2009, December 22) from http://www.linkedin.com/news?viewArticle=&articleID=95616999&gid=96032&articleURL=http%3A%2F%2Fwww.slapapp.com%2Ffree-iphone-promo-codes-for-parkn-find-navigation-iphone-app&urlhash=794g&trk=news_discuss

BANKING

New mastercard service simplifies the search for atms by text messaging atm location to cardholder's cell phone. (n.d.). Retrieved from http://www.sybase.com/detail?id=1050432

Bills, SB. (2009, June 25). Boston iso offers blackberry app. American Banker, American Banker. New York, N.Y.: Jun 25, 2009. Vol. 174, Iss. 121; pg. 9.

Anonymous, Initials. (2009, July 21). Cash carrying consumers look to prepaid debit cards to avoid credit card woes . PR Newswire, Retrieved from http://ezproxy.spl.org:2048/login?url=http://proquest.umi.com/pqdweb?did=1797813891&Fmt=3&clientId=11206&RQT=309&VName=PQD doi: 1797813891

Anonymous, Initials. (2009, November 3). Sybase 365 iphone mobile banking application enables financial institutions to further extend full-service mobile banking to their customers anywhere in the world. Business Wire, (1892286461)

https://www.usaa.com/inet/ent_utils/McStaticPages?key=usaa_mobile_iphone_main
\Anonymous, Initials. (2009, July 13). Video: pnc virtual wallet(sm) student gives parents peace of mind, helps college students make the grade with their finances:spending tracker, reimbursement request feature and e-mail alerts to parents empower students to better manage their money. Retrieved from http://ezproxy.spl.org:2048/login?url=http://proquest.umi.com/pqdweb?did=1784539701&Fmt=3&clientId=11206&RQT=309&VName=PQD doi: 1784539701
Anonymous, Initials. (2009, July 21). Visa europe: visa gears up

to offer real-time mobile alerts; new visa service keeps consumers informed of their visa card activity via mobile . M2 Presswire, Retrieved from http://ezproxy.spl.org:2048/login?url=http://proquest. umi.com/pqdweb?did=1797124931&Fmt=3&clientId=11206&R QT=309&VName=PQD doi: 1797124931.

Anonymous, Initials. (2009). * concur mobile for iphone now available on apple's app store:free mobile application enables concur travel & expense clients to easily change or update itineraries, capture expenses and approve expense reports - all from their iphone . PR Newswire, Retrieved from http://ezproxy.spl.org:2048/ login?url=http://proquest.umi.com/pqdweb?did=1891896091&Fm t=3&clientId=11206&RQT=309&VName=PQD doi: 1891896091

BUDGETING

Anonymous, Initials. (2007, August 6). Yodlee launches yodlee mobile sms. Business Wire.

INSURANCE

Back to school theft alert: allstate identifies top five stolen items and how to protect them . (2006). Business Wire, Retrieved from http://ezproxy.spl.org:2048/login?url=http://proquest.umi.com/pqd web?did=1092424431&Fmt=3&clientId=11206&RQT=309&VNa me=PQD doi: 1092424431

R Raphael, JRR. (2009, December 15). Android app alert: mobile defense hits the market. PC World, 184739, Retrieved from http:// www.pcworld.com/article/184739/android_app_alert_mobile_defense_hits_the_market.html

CONNECTIVITY

Nguyen, BN. (2009). Get the new Boingo W-Fi software for Android!. Linkedin. Retrieved (2009, December 22) from http://www. linkedin.com/groupAnswers?viewQuestionAndAnswers=&discussion ID=11055257&gid=39723&trk=EML_anet_qa_ttle-d7hOonOJum-NFomgJt7dBpSBA

Nicholson, DN. (2009, December 10). Try gogo inflight internet for free. Retrieved from http://www.linkedin.com/newsArticle?viewDis cussion=&articleID=93391797&gid=56468&trk=EML_anet_nws_

Anonymous, Initials. (2008, December 18). 2009 inauguration mobile app launched by patton boggs and qorvis. PR Newswire, Retrieved from http://ezproxy.spl.org:2048/login?url=http://proquest.umi.com/pqdweb?did=1614136811&Fmt=3&clientId=11206&RQT=309&VName=PQD doi: 1614136811

Anonymous, Initials. (2009, December 3). Award-winning samsung mondi now available in dallas/fort worth market to support rollout of mobile wimax service. Business Wire.

GPS

\Broida, RB. (2010, January 8). 'when's the next starbucks?' iexit app lists freeway pois. Retrieved from http://reviews.cnet.com/iphone-atlas/?keyword=Drive

DiGregorio, MD. (2009). LifeInPocket(R) Rolls Out World's Most Powerful App for BlackBerry: Offers the Capability of Hundreds of Advanced Apps While Remaining Small, Fast, Fully Personalized, Powerful and Free. Entrepreneur. Retrieved (2009, December 22) from http://www.entrepreneur.com/prnewswire/release/221663.html

MAPS

Anonymous, Initials. (2008). Navigon provides high-end features at entry-level prices with the launch of the 2100 max and 2120 max gps devices:new directhelp feature provides instant links to nearby hospitals and roadside assistance. PR Newswire, Retrieved from http://ezproxy.spl.org:2048/login?url=http://proquest.umi.com/pqdweb?did=1462194971&Fmt=3&clientId=11206&RQT=309&VName=PQD doi: 1462194971

Mann, MM. (2009). 13 mobile apps that work - presentation transcript. Retrieved from http://www.slideshare.net/marinamann/13-mobile-apps-that-work

Anonymous, Initials. (2009, October 13). Inrix selected by south carolina department of transportation to provide real-time traffic speeds and travel times statewide:1,200 miles of roadway coverage added through the highly successful i-95 corridor coalition vehicle probe project. PR Newswire, Retrieved from http://ezproxy.spl.org:2048/login?url=http://proquest.umi.com/pqdweb?did=1878034581&Fmt=3&clientId=11206&RQT=309&VName=PQD doi: 1878034581

ROADSIDE ASSISTANCE
Anonymous, Initials. (2008). Metropolitan transportation commission; bay area's 511 phone service adds freeway assistance option. Science Letter, Retrieved from http://ezproxy.spl.org:2048/login?url=http://proquest.umi.com/pqdweb?did=1556899071&Fmt=3&clientId=11206&RQT=309&VName=PQD doi: 1556899071

AIRLINES
Anonymous, Initials. (2009). American airlines: american airlines makes it easier than ever to book en espanol at aa.com; popular "search by price & schedule" option now available in spanish. M2 Presswire, Retrieved from http://ezproxy.spl.org:2048/login?url=http://proquest.umi.com/pqdweb?did=1642454251&Fmt=3&clientId=11206&RQT=309&VName=PQD doi: 1642454251

New! mobile boarding passes. (n.d.). Retrieved from http://www.united.com/page/article/0,6867,66,00.html?navSource=RelatedLinks.

FLIGHT/HOTEL
Amon Cohen, AC. (2009, November 23). Booker hotelzon to expand mobile service. Business Travel News, 26(17)

Neman, SN. (2009, May 19). Hyatt launches a very useful mobile website. Retrieved from http://blog.clubtexting.com/2009/05/hyatt-launches-a-very-useful-mobile-website.html

TRANSPORTATION
Designated driver iphone app - reach home safely. (2009, July 15). Retrieved from http://www.ziphone.org/2009/07/ensure-safe-drive-with-designated.html Isixt 3.0 car hire app unleashed. (2009, October 12). Retrieved from http://www.mobiletor.com/2009/10/12/isixt-3-0-car-hire-app-unleashed

Carpooling startups avego, icarpool and zimride open the door for community rideshares. (2009, April 12). Retrieved from http://www. digitalverdure.com/1/post/2009/04/carpooling-startups-zimcar-icarpool-and-avego-take-ridesharing-to-the-next-level.html

INSURANCE

Anonymous, Initials. (2009). Global travelers embrace hth world-wide's mpassport(r); mobile medical concierge builds global user base with popular iphone app. PR Newswire, doi: 1918557431

Hth worldwide introduces trip protection insurance with major medical benefits. (2009). PR Newswire, Retrieved from http://ezproxy.spl. org:2048/login?url=http://proquest.umi.com/pqdweb?did=186340431 1&Fmt=3&clientId=11206&RQT=309&VName=PQD
doi: 1863404311

ALERTS

Anonymous, Initials. (2009). Eyeline communications inc; arrivedok flight arrival tracker lets air passengers notify others about their personal landing by sms, email, blogs, twitter. Science Letter, (15389111), Retrieved from http://ezproxy.spl.org:2048/login?url=http://proquest.umi.com/pqdweb?did=1671774531&Fmt=3&clientId=11206&RQT=309&VName=PQD doi: 1671774531

Anonymous, Initials. (2009, August 12). videonext launches free ip surveillance software on the iphone:cavu, cavu pro and cavu free now available on apple's app store . PR Newswire, 1827022261.

Sekhar, PS. (2007, August 4). Info-tech: alerts on sms about house break-in, car theft!. Businessline, Retrieved from http://ezproxy.spl.org:2048/login?url=http://proquest.umi.com/pqdweb?did=1315822571&Fmt=3&clientId=11206&RQT=309&VName=PQD doi: 1315822571
Orbitz asks via online chat: can i help you?; online travel site begins beta test of 'orbitztlc live chat' support for customers searching for vacation packages. (2006, October 11). PR Newswire, Retrieved from http://ez-proxy.spl.org:2048/login?url=http://proquest.umi.com/pqdweb?did=1 143790121&Fmt=3&clientId=11206&RQT=309&VName=PQD doi: 1143790121

Mann, MM. (2009). 13 mobile apps that work - presentation transcript. Retrieved from http://www.slideshare.net/marinamann/13-mobile-apps-that-work

SAFETY

Grobart, SG. (2009). For Drivers who can't resist calling, a high-tech nanny to cut off phones:[series] . New York Times, Retrieved from http://ezproxy.spl.org:2048/login?url=http://proquest.umi.com/pqd web?did=1905808051&Fmt=3&clientId=11206&RQT=309&VNa me=PQD doi: 1905808051

Weisbaum, HW. (2006). How often should Motor oil be changed?. msnbc.com, Retrieved from http://www.msnbc.msn.com/ id/12359794/ns/business-consumer_news

Ashe, SA. (2009, December 10). Need a new headlight bulb? there's an app for that. Retrieved from http://reviews.cnet.com/ iphone-atlas/?keyword=apps

Anonymous, Initials. (2008). New solution for campus safety pushes emergency alerts to scrolling led displays for better reach to visitors, students and staff:inova solutions led wallboard joins e2campus network to increase campus safety. PR Newswire, Retrieved from http://ezproxy.spl.org:2048/login?url=http://proquest.umi.com/pqd web?did=1509829201&Fmt=3&clientId=11206&RQT=309&VNa me=PQD doi: 1509829201

Anonymous, Initials. (2009, June 1). K-12 schools select e2campus for parent notification system . PR Newswire.

Broadcast alert service introduced by redsky. (2007, July 31). Telecomworldwire, (1312495911).

SHAMAH, DS. (2009, November 27). Hang up the car keys. Jerusalem Post, Retrieved from http://ezproxy.spl.org:2048/ login?url=http://proquest.umi.com.ezproxy.spl.org:2048/pqdweb?di d=1917126141&sid=1&Fmt=3&clientId=11206&RQT=309&VNa me=PQD doi: 1917126141

DIRECTORIES

Anonymous, Initials. (2008, July 7). jingle networks, inc.; tnt pro-motes nba playoffs via 1-800-free411. Jingle Networks, Inc.; TNT Promotes NBA Playoffs via 1-800-FREE411.
The Charleston Gazette, Retrieved from http://ezproxy.spl.org:2048/login?url=http://proquest.umi.com/pqdweb?did=1633907631&Fmt=3&clientId=11206&RQT=309&VName=PQD doi: 1633907631

Anonymous, Initials. (2009, October 13). privus mobile debuts on win-dows mobile 6.5, releases free caller name lookup upgrade for windows mobile app. Business Wire, Retrieved from http://ezproxy.spl.org:2048/login?url=http://proquest.umi.com/pqdweb?did=1878708971&Fmt=3&clientId=11206&RQT=309&VName=PQD doi: 1878708971

WEB

Microsoft bing comes to the iphone. (2009, December 16). Retrieved from http://topiphonenews.com/microsoft-bing-comes-to-the-iphone

ROW, Initials. (2009). RSS Twitter Facebook Subscribe to MakeUseOf. Now, 140007 members! IQMobileSearch: All-In-One Search Engine for iPhone. Iqmobilesearch. Retrieved (2009, De-cember 22) from http://www.makeuseof.com/dir/iqmobilesearch-search-popular-sites

MANUALS & INSTRUCTIONS

MARTIN COOPER - INVETOR OF THE CELL PHONE. (N.D.). RETRIEVED FROM HTTP://WWW.CELLULAR.CO.ZA/CELLPHONE_INVENTOR.HTM.

FOOD

Anonymous, Initials. (2009, November 17). Number one rule of gift giving this year: shop for quality not quantity:trusted brands and superior products should be top of list for shoppers . PR Newswire.

LAW

DICTIONARY OF LEGAL AND LAW TERMS. (N.D.). RETRIEVED FROM HTTP://APPSHOPPER.COM/REFERENCE/DICTIONARY-OF-LEGAL-AND-LAW-TERMS

Hill, KH. (2009, May 18). Top ten iphone apps for biglawyers. Retrieved from http://abovethelaw.com/2009/05/lawyers_and_iphone_apps_draft.php

Lurssen, AL. (2009, October 14). jd supra's 'legal edge' iphone app allows lawyers to connect with mobile users. Retrieved from http://scoop.jdsupra.com/2009/10/articles/content-marketing/jd-supras-legal-edge-iphone-app-allows-lawyers-to-connect-with-mobile-users

Ut law iphone web app. (2009, September 24). Retrieved from http://www.utexas.edu/law/m/about.html

ANIMALS

Lostmydoggie.com. (n.d.). Retrieved from http://www.lostmydoggie.com/?gclid=CKKH3O3WjqACFQIhDQodNRTEeg

Shreve, JS. (2005, December 06). Petsmobility. Retrieved from http://www.petsmobility.com/products/pawTrax.php

RESOURCES

Pedidoser:a easy to use reference guide for pediatric outpatients, available for android, palm, iphone and more. (2009, July 30). Retrieved from http://www.androidmedapps.com/index.php/2009/07/pedidosera-easy-to-use-reference-guide-for-pediatric-outpatients-available-for-android-palm-iphone-and-more

Shupe, AS. (2008, January 25). Two nanny agencies to adopt groundbreaking software to utilize in child care. Retrieved from http://www.business-opportunities.biz/2008/01/25/two-nanny-agencies-to-adopt-groundbreaking-software-to-utilize-in-child-care

Your mobile phone can save you in emergency (general tips). (2008, April). Retrieved from http://forums.techarena.in/off-topic-chat/1135337.htm

MONITORING

Belic, DB. (2009, December 2). Net nanny, smobile systems team up to launch net nanny mobile read more: http://www.intomobile.

com/2009/12/02/net-nanny-smobile-systems-team-up-to-launch-net-nanny-mobile.html#ixzz0b8otdjsp. Intomobile, Retrieved from http://www.intomobile.com/2009/12/02/net-nanny-smobile-systems-team-up-to-launch-net-nanny-mobile.html

Arpita, Initials. (2009, June 22). phone creeper for windows mobile - spy app for windows mobile. Retrieved from http://www.us.ayushveda.com/phone-creeper-spy-app-for-windows-mobile

ALERTS
Amber alert iphone app goes multiplatform. (2009, March 11). Retrieved from http://blog.clubtexting.com/2009/03/amber-alert-iphone-app-goes-multiplatform.html

FUN
Lisa Schuster, LS. (2009, November 17). Santa goes green by going mobile to benefit the march of dimes. Press Release Central.
Hartstein, JH. (2008, November 26). Keys please: drunk dial a designated driver. Retrieved from http://www.culture-buzz.com/blog/Keys-Please-Drunk-Dial-a-Designated-Driver-1938.html

ADDITIONAL TIPS
Cell phone glossary - cell phones - mobiledia. (n.d.). Retrieved from http://www.mobiledia.com

Glossary of cellular terms. (2006, January 25). Retrieved from http://www.phonedog.com/cell-phone-buying-guide/glossary-of-cellular-terms.aspx

Webopedia: online computer dictionary for computer and internet terms and definitions. (n.d.). Retrieved from http://www.webopedia.com

CONTACT INFORMATION

If you would like to obtain additional services you may contact any of the following resources.

Denise Barnes
iKeep It Funky
Mobile Cell Phone Assistance & Information
Tel: 206-659-8239
Email: Info@iKeepItFunky.com
Web: http://www.iKeepItFunky.com

Rob Ripley
Contract Graphic & Web Designer
Email: designer.ripley@gmail.com
Web: http://www.ripleysite.net

Mélanie Hope
StormKatt Productions
Editing, writing, design
Email: StormKatt@gmail.com
Web: http://StormKatt.com
PO Box 434 Kent, WA 98035
253-208-5901

Marrico Gordon
Writer And Poet
Email: marrico.gordon@yahoo.com
206-679-9786

Rick Farrington
Rebecca Little - Marketing Director
Mobile Marketing Solutions for the 21st Century!
206-372-4445
Email: Rick@optinmobilemarketingsolutions.com
Email: Rebecca@optinvip.net
Web: http://www.optinvip.com
Web: http://www.mcoups.com

www.ingramcontent.com/pod-product-compliance
Lightning Source LLC
Chambersburg PA
CBHW071427170526
45165CB00001B/430